SCI PUBLICATION 244

Protection of Buildings against Explosions

E YANDZIO   BSc, MEng, CEng, MIMarE
M GOUGH   BEng, MSc, DIC

Published by:

The Steel Construction Institute
Silwood Park
Ascot
Berkshire SL5 7QN

Tel: 01344 623345
Fax: 01344 622944

© 1999 The Steel Construction Institute

Apart from any fair dealing for the purposes of research or private study or criticism or review, as permitted under the Copyright Designs and Patents Act, 1988, this publication may not be reproduced, stored or transmitted, in any form or by any means, without the prior permission in writing of the publishers, or in the case of reprographic reproduction only in accordance with the terms of the licences issued by the UK Copyright Licensing Agency, or in accordance with the terms of licences issued by the appropriate Reproduction Rights Organisation outside the UK.

Enquiries concerning reproduction outside the terms stated here should be sent to the publishers, The Steel Construction Institute, at the address given on the title page.

Although care has been taken to ensure, to the best of our knowledge, that all data and information contained herein are accurate to the extent that they relate to either matters of fact or accepted practice or matters of opinion at the time of publication, The Steel Construction Institute, the authors and the reviewers assume no responsibility for any errors in or misinterpretations of such data and/or information or any loss or damage arising from or related to their use.

Publications supplied to the Members of the Institute at a discount are not for resale by them.

Publication Number: SCI-P-244

ISBN 1 85942 089 3

British Library Cataloguing-in-Publication Data.
A catalogue record for this book is available from the British Library.

FOREWORD

The threat to commercial property and personnel from attack is regrettably a reality and the scale of terrorism has increased worldwide. This publication provides guidance on the protection of commercial property and personnel from the effects of explosions caused by the detonation of high explosives. It will be of assistance to engineers and architects who are involved in building designs where this type of protection is required.

Particular thanks are due to the following for their assistance:

Mr C Bowes	TPS Consult
Mr C J R Veale	Government Security Advisor
Mr F Hulton and Major J F Mackenzie	Defence Evaluation and Research Agency
Mr A Mann	Allott and Lomax
Mr G Fairlie	Century Dynamics

Funding was provided by

- The Construction Innovation and Research Management Division of the DETR Construction Sponsorship Directorate under its Partners in Technology programme

- The Steel Construction Industry Federation (SCIF)

- British Steel plc.

CONTENTS

Page No.

SUMMARY		viii
1	INTRODUCTION	1
2	BOMB DAMAGE TO BUILDINGS	2
2.1	World Trade Center, New York City	2
2.2	Murrah Federal Building, Oklahoma City	3
2.3	St Mary Axe, London	3
2.4	Manchester City Centre	4
2.5	London Docklands	4
2.6	Staples Corner, North London	6
2.7	Jewish Community Centre, Buenos Aires, Argentina	6
3	PROTECTION PHILOSOPHY FOR BUILDINGS	7
3.1	Protection by design	7
4	PREVENTIVE MEASURES	12
4.1	External layout planning	12
4.2	Access control	14
4.3	Bomb shelter areas	15
4.4	Management procedures	15
5	BUILDING DESIGN	16
5.1	Requirements of the Building Regulations	16
5.2	Requirements of the Codes of Practice	17
5.3	Guidance documents and technical manuals	17
5.4	Choice of structural system	18
6	EXPLOSIVES	21
6.1	Sources of explosion	21
6.2	Explosive materials	21
6.3	Explosion process for high explosives	22
6.4	Explosive devices and methods of delivery	22
6.5	TNT equivalency	24
7	BLAST OVERPRESSURE AND LOADS	26
7.1	Types of explosion	26
7.2	Scaled distance	29
7.3	Blast wave characteristics in free air	30
7.4	Reflected blast wave characteristics	31

7.5	Blast overpressures due to a surface burst	35
7.6	Mathematical representation of blast overpressure profile	35
7.7	Blast pressures and loads on buildings	37
7.8	Internal explosions in buildings	41
7.9	Computational methods for predicting blast overpressure and loads	42

8 DYNAMIC RESPONSE — 44
- 8.1 Response regimes — 44
- 8.2 Analysis methods — 45
- 8.3 Single-degree-of-freedom method — 47
- 8.4 Multiple-degree-of-freedom method — 51
- 8.5 Energy methods — 53
- 8.6 Pressure-impulse diagrams — 54

9 MATERIAL PROPERTIES — 59
- 9.1 Behaviour of materials under dynamic loading — 59
- 9.2 Strain-rate effects in structural steels — 59
- 9.3 Strain-rate effects in stainless steels — 61
- 9.4 Strain-rate effects in concrete — 62

10 STRUCTURAL DESIGN APPROACH — 63
- 10.1 Design requirements — 63
- 10.2 Acceptance criteria — 67
- 10.3 Procedure to assess response and adequacy of structural components — 70

11 DETAIL DESIGN AND STRUCTURAL CONNECTIONS — 72
- 11.1 Beams — 72
- 11.2 Columns — 72
- 11.3 Floors — 73
- 11.4 Roofs — 73
- 11.5 Walls — 73
- 11.6 Cladding — 74
- 11.7 Stairs — 74
- 11.8 Beam-column connections — 74

12 FOUNDATIONS — 77

13 NON-STRUCTURAL ENHANCEMENTS — 78
- 13.1 Blast-enhanced glazing — 78
- 13.2 Facade detailing — 79
- 13.3 Internal layout of building — 80

14	INSPECTION OF DAMAGED BUILDINGS	81
REFERENCES		82
APPENDIX A	BOMB SHELTER AREAS	89
APPENDIX B	BLAST WAVE PARAMETERS	91
APPENDIX C	TRANSFORMATION FACTOR TABLES AND RESPONSE CHARTS	96
APPENDIX D	DESIGN FOR GLAZING PROTECTION	103

SUMMARY

This publication provides guidance on the design of commercial and public buildings where there is a requirement to provide protection against the effects of explosions caused by the detonation of high explosives.

A philosophy for the design of buildings to reduce the effects of attack is introduced and a design procedure is proposed. The steps in the procedure cover threat assessment, preventive measures, the evaluation of blast overpressure and loads, and the response and adequacy of structural components subjected to blast. The background and basis of each of the steps is explained.

The *robustness* of buildings and the prevention of disproportionate collapse are discussed in relation to the UK Building Regulations and Codes of Practice, and to technical guidance documents that have been developed by military establishments. Enhancements to structural frames and to non-structural components are proposed, and the inspection of buildings damaged by blast loads is considered.

A comprehensive list of references is provided.

Resumé

Protection des immeubles contre les explosions

Cette publication est destinée à servir de guide de dimensionnement d'immeubles publics ou commerciaux où il est nécessaire de prévoir une protection contre les effets d'explosions causées par des explosifs.

Une philosophie de dimensionnement des immeubles pour réduire les effets d'une attaque est développée et une procédure de dimensionnement est proposée. Les étapes de cette procédure couvrent la prise en compte des menaces, les mesures de prévention, l'évaluation des surpressions et des charges lors de l'explosion, et la réponse et l'adéquation des composantes structurales soumises à l'explosion. Les bases de chaque étape sont expliquées.

La robustesse des immeubles et la prévention d'écroulements disproportionnés sont discutés en relation avec les codes UK de construction et les documents de guidance technique développés pour les sites militaires. Des améliorations des composantes structurales et non structurales sont proposées. L'estimation des dégâts d'un immeuble endommagé par une explosion est également abordée.

La publication comporte également une liste bibliographique.

Schutz von Gebäuden gegen Explosionen

Zusammenfassung

Diese Publikation gibt eine Anleitung für die Berechnung von Geschäftshäusern und Öffentlichen Bauten bei denen Anforderungen zum Schutz gegen Explosionen infolge Sprengstoff bestehen.

Eine Philosophie zur Berechnung von Gebäuden, mit dem Ziel die Auswirkungen eines Angriffs zu reduzieren, wird vorgestellt, sowie ein Berechnungsverfahren vorgeschlagen. Die einzelnen Schritte behandeln die Abschätzung der Bedrohung, vorbeugende Maßnahmen, die Beurteilung des Explosionsdrucks und der Lasten sowie Antwort und Angemessenheit der Bauteile unter Explosionsdruck. Der Hintergrund und die Grundlagen für jeden einzelnen Schritt werden erklärt.

Die Widerstandsfähigkeit von Gebäuden und die Vermeidung eines unverhältnismäßigen Versagens werden besprochen, mit Bezug auf die englischen Bauvorschriften und Normen sowie Dokumente, die von militärischen Einrichtungen erstellt wurden. Verbesserungen für Tragwerke und nichttragende Elemente werden vorgeschlagen sowie die Inspektion von durch Explosionslasten beschädigten Gebäuden berücksichtigt.

Eine umfassende Referenzliste wird bereitgestellt.

Protezione degli edifici contro le esposioni

Sommario

Questa pubblicazione è una guida alla progettazione di edifici commerciali e pubblici per i quali viene esplicitamente richiesta la protezione contro le esplosioni causate dalla detonazione di potenti esplosivi.

Viene introdotta la filosofia progettuale relativa alla riduzione degli effetti delle esplosioni sugli edifici, ed è presentata la metodologia di progetto associata. Le fasi applicative di questa procedura riguardano la valutazione di minacce, le misure di prevenzione, la stima delle sovrapressioni di scoppio e dei carichi, e la risposta e l'adeguatezza delle componenti strutturali interessate dallo scoppio. Sono fornite le principali nozioni teoriche e le basi associate ad ogni fase della procedura.

La robustezza dell'edificio e la prevenzione di collassi di rilevanti proporzioni è discussa con riferimento sia al Regolamento sugli Edifici del Regno Unito sia al vigente Codice Progettuale. Sono inoltre considerati i documenti tecnici sviluppati da specifici organismi militari. Vengono proposte tecniche di miglioramento dei sistemi strutturali intelaiati e delle componenti non strutturali e viene trattata l'ispezione di edifici danneggiati da esplosioni.

In aggiunta, nella pubblicazione viene riportato un'esaustivo elenco di riferimenti relativi alla tematica trattata.

Protección de edificios frente a explosiones

Resumen

Esta publicación es una guia para el proyecto de edificios públicos y comerciales donde se plantea como requisito la protección contra presiones causadas por detonaciones do explosivos.

Se plantea tanto una filosofía como un método de proyecto para reducir los efectos del ataque. Los pasos del método cubren la cuantificación de la amenaza, medidas preventivas, cálculo de la sobrepresión de explosión y cargas, y la adecuación y respuestas de las componentes esructurales sometidas a la detonación.

Figure 2.2 *Cladding damage to a commercial building, City of London [Photo courtesy of TPS Consult]*

to be crazed. The 33-mm thick laminated glass windows at street level survived without crazing.

2.4 Manchester City Centre

In June 1996, a bomb was detonated in the Arndale Centre in central Manchester, causing extensive damage (Figure 2.3); many injuries were caused by flying glass. Following preliminary structural assessments, it was found that damage caused by the explosion was due mainly to the destruction of glazing and cladding panels and, although glazing damage was extensive, it appeared random. Ground-floor windows relatively close to the blast remained intact, while windows much further away and at high elevation were shattered. Where structural damage did occur, it was relatively light. The worst case of structural damage occurred near the heart of the explosion, where the structural frame of a 200-tonne pedestrian bridge was twisted and lifted off its bearings, leaving it dangerously unsafe.

2.5 London Docklands

A terrorist bomb was detonated in London Docklands in February 1996. The device consisted of home-made explosives and was placed in a flatbed truck. Two people were killed and office buildings and nearby homes were damaged extensively. The composite concrete-steel viaduct of the Docklands Light Railway was also damaged, although damage was restricted to peripheral items and the box girder nearest the blast suffered only minimal damage. There was little structural damage to surrounding buildings, however glazing and cladding damage was extensive: no glazing survived within 50 m of the blast (Figure 2.4).

Figure 2.3 *Damage to a commercial building showing deformed framing [Photo courtesy of TPS Consult]*

Figure 2.4 *Extensive damage to cladding after explosion, London Docklands [Photo courtesy of TPS Consult]*

2.6 Staples Corner, North London

In 1992, there was an explosion at Staples Corner, London, as a result of which a single-storey, steel-framed warehouse was damaged. The explosion occurred approximately 17 m from the nearest corner of the building. The walls and roof were clad with light-profiled steel sheeting supported on purlins. As the cladding was of frangible construction, severe damage was limited to the cladding, its supporting steelwork and connections. Only minimal damage was sustained by the main steel frame.

2.7 Jewish Community Centre, Buenos Aires, Argentina

The Jewish Community Centre in Buenos Aires was located in a densely-constructed area. In 1994, an explosive was detonated approximately 4 m from the building. The exterior walls of the five-storey building had been constructed using masonry brickwork, which also supported the floor slabs. The bomb blast virtually destroyed the whole building. The exterior walls were demolished completely by the blast and the floor slabs collapsed progressively.

3 PROTECTION PHILOSOPHY FOR BUILDINGS

This Section introduces a philosophy for the protection of commercial buildings and their occupants from the effects of explosions. Many of the buildings standing today have been built with minimum consideration of protection for self-defence purposes. Without a change of policy and greater awareness, new buildings will be constructed likewise.

Protection of a building is the aggregate sum of all the measures that can be employed to ensure that the threat from explosions is made ineffective. For the protection to be effective, the building should be capable of:

- deflecting an attack by showing, through layout, security, and defence, that the chance of success for the terrorist is small
- disguising the valuable and vulnerable parts of a potential target, so that the emphasis of the attack is focused on the wrong area
- dispersing a potential target, so that an attack would not cover an area large enough to cause significant destruction
- stopping an attack from reaching a potential target, by erecting a physical barrier to the method of attack
- minimising the effect of the attack once its reaches the target by strengthening the structure.

Protecting buildings from explosions does not just mean protecting people and property; it includes protecting the business. The ability of an organisation to recover quickly, no matter what the disaster, is becoming essential if it is to survive an attack. The protection of key assets may therefore be critical in pursuing this aim successfully. A threat assessment will define the requirements specifically, and point to where the emphasis in counter-measures should lie.

3.1 Protection by design

To achieve protection by design, the building must be designed to:

- minimise the likelihood and magnitude of threat (by adopting preventive measures that discourage or impede an attack)
- prevent collapse; where collapse occurs, it must be local rather than disproportionate or catastrophic; substantial damage short of total collapse may be acceptable based on the design premise
- prevent fire; where fire occurs, it must not burn out of control
- protect the people and assets from the effect of blast waves and projectiles (i.e. glass, bricks, etc.)
- provide the occupants of the building with either a safe area or an effective escape route and assembly area
- enable repair to be performed effectively after an attack.

The implementation of the above philosophy in the design of the building is best presented in the form of a flowchart (Figure 3.1). The flowchart shows the activities and their sequence such that an effective scheme for protecting people, property, and the business can be developed.

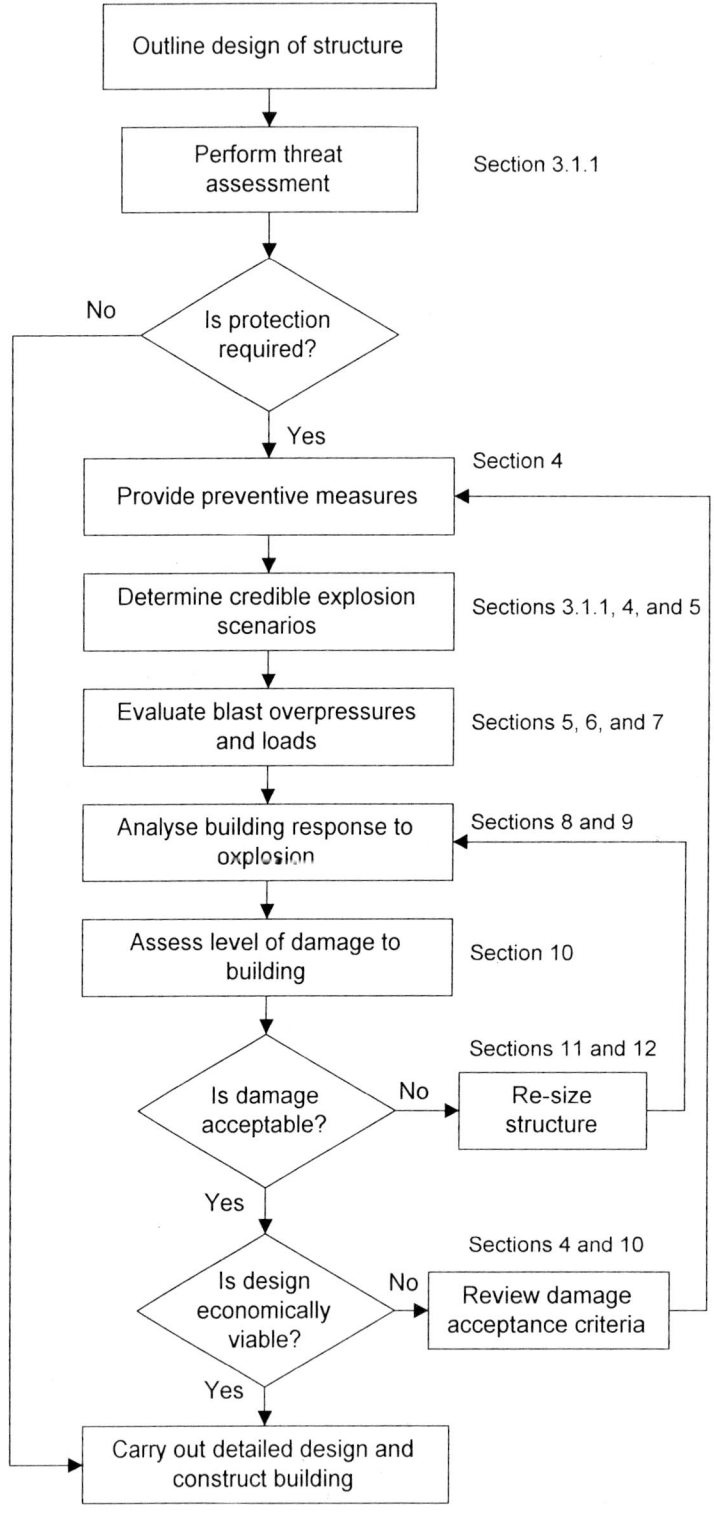

Figure 3.1 *Flowchart showing the philosophy for protecting buildings against explosions*

The protection provided for commercial buildings, occupants, and equipment will need to satisfy the requirements of the following regulations:

- The Management of Health and Safety at Work Regulations 1992[1]
- The Construction Products Directive of the European Communities, Annex 1[2]
- The Construction (Design and Management) Regulations 1994[3].

These regulations legally oblige a building owner or employer:

- to provide safety to employees when there is a threat of a terrorist act
- to ensure that equipment is fit for its intended use
- to ensure that risk management throughout the life-cycle is established.

The following Sections discuss briefly the activities that should be undertaken initially to choose the most appropriate options. These options will be developed in the main design.

3.1.1 Threat assessment

A threat assessment is a risk analysis in which all the elements that could constitute a threat are identified and evaluated, together with the degree of probability that specific risks will occur, and their potential consequences.

The probability or likelihood of an attack on a building is determined by:

- the degree of social and political stability in the country
- whether there is a history of terrorism
- the nature of the business of the owner or occupier
- the building's location (is it a prestige location?)
- the building's proximity to neighbouring buildings that might be possible targets.

To determine the nature of the threat, access must be available to relevant statistics about attacks in similar locations or circumstances. These statistics should contain as much information as possible, including as a minimum the date, time, type, size, location, and reasons why the attack took place. Ideally, to aid in the threat assessment, a database should be compiled where all the information can be recorded. Professional advice can be sought from public agencies, such as the Police and Home Office, and from private organisations or from the publication *Bombs - protecting people and property*[4]. The alternative is to research local media sources, however this would be time consuming and less effective.

The magnitude of the threat depends to some extent on the size of the explosive device that can be delivered. Discussion about devices is presented in Section 6.4.

The potential consequences of an attack that need to be considered include:
- human casualties - death, injury (physical and psychological)
- reconstruction costs
- temporary relocation costs
- loss in trade
- loss in man hours
- loss of key employees
- reduction in workforce motivation
- legal action.

The assessment process should be sufficient to provide the basis for the design of the building, however threat assessments should be made at intervals during the life of the building to reflect contemporary circumstances and, if necessary, to initiate modifications to improve protection.

3.1.2 Preventive measures

Preventive measures will play an important role in mitigating the effects of a terrorist attack. Practical steps to reduce the magnitude of the attack should be established during the initial design stages of the project. These measures are discussed in detail in Section 4.

3.1.3 Determine credible explosion scenarios

Based on the preventive measures that have been chosen (see Section 4) and the outcome of the threat assessment studies (see Section 3.1.1), credible explosion scenarios should be identified. Information will include the type, size, and location of the explosive source.

3.1.4 Evaluate blast overpressure and loads

Blast overpressure characteristics that will define the loading to be used in the subsequent response analyses are presented in Section 7. Both hand calculations and computational methods are described. Section 7 also provides information on the determination of loading on solid buildings.

3.1.5 Analyse building response to explosion

The response of the building to the loading identified in Section 7 can be based on one or more of the analysis methods proposed in Section 8. Both simplistic and more sophisticated numerical analyses are considered. Pressure-impulse diagrams are introduced as a simple but effective way to study the various response regimes that may occur.

3.1.6 Assess level of damage to building

The level of damage to the building can be assessed on the basis of the building design philosophy in Section 5, the material property information in Section 9, and the structural design methodology proposed in Section 10. Acceptance for limits to allowable damage may be based on strength and deformation criteria. A convenient form of representation of damage for explosion scenarios can be obtained from pressure-impulse diagrams (see Section 8.5).

Considerations for detail design of components of the building and structural connections are presented in Section 11.

The resistance of building foundations against blast loading are considered in Section 12.

Non-structural enhancements to provide significant additional protection to the building are presented in Section 13. These include providing blast-enhanced glazing, considering facade detailing such that blast overpressures are minimised, and considering the internal layout of the building such that injury to people and damage to assets are minimised.

3.1.7 Economic viability

Protection against terrorism can never be completely successful and there is clearly a level of protection where the cost of protection with respect to the cost of loss is optimised. It is very difficult to estimate the cost of protection as it depends on the extent of protection that is to be provided. Overall building costs for some projects have resulted in minimal increases, i.e. 3-5%, while cost increases have been significant for other projects.

4 PREVENTIVE MEASURES

In many cases, the cheapest method of securing protection against the effects of blast is to adopt preventive measures. These generally include:

- external layout planning
- access control (security, closed circuit television)
- management procedures.

Measures appropriate to the assessed threat should be considered fully when a new structure design is to be undertaken or where refurbishment of existing structures takes place. In many cases, however, the best preventive measures may not be practical, for example important buildings generally need to be located in cities for functional or operational reasons and this limits influence over external layout.

4.1 External layout planning

When the external layout around the building is being planned, there are many aspects that should be considered to minimise damage from explosions. Figure 4.1, taken from *Blast and ballistic loading of structures*[5], shows a possible external layout, taking into account aspects discussed below.

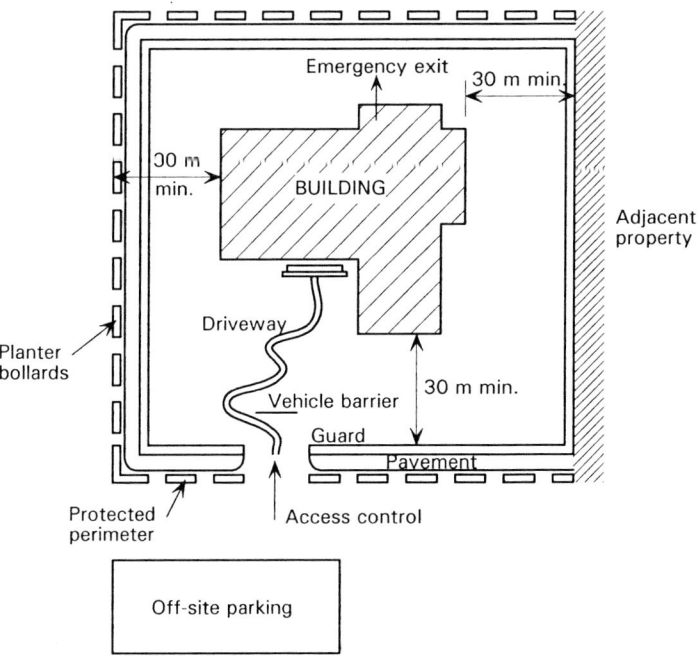

Figure 4.1 *Schematic layout of site for protection against bombs*

Landscaping

One of the important measures in minimising the effects of a device on both people and property is the creation of distance (stand-off), i.e. creating measures that force the device to be placed as far away from the building as possible.

To maximise the stand-off from a vehicle-borne device, for example by preventing vehicles from mounting the pavement and parking next to the building or approaching the front entrance, the installation of bollards and other items of street furniture should be considered. Hard landscaping incorporating steps and mounds will also contribute to creating stand-off. Care should be taken if considering soft landscaping, however, to ensure that concealment places for smaller, portable devices are not created in dense planting, etc. Buildings in dips are not usually favoured.

Perimeter protection

Where the location of the site and local planning conditions permit, it will be normal to provide a physical barrier (a fence or wall) at or near the site boundaries. This would provide a stand-off distance between the building and public highway on which a vehicle bomb might be positioned. In order to provide effective blast protection, the barrier must be sited, detailed, and protected carefully. It would also be beneficial to include a perimeter intruder-detection system to alert security personnel of any potential threat.

Roads and railways

The site should preferably not be adjacent to or divided by any raised viaduct carrying a road, railway, canal, cycle track, or other public right of way.

Rights of way

Sites with rights of way crossing them should be avoided. Rights of way can sometimes be extinguished or diverted but this can involve a lengthy and expensive legal process. Natural short-cuts across a site may need to be considered in the context of the neighbourhood. Other planned developments may change the pattern of movement and provide pedestrians with more attractive alternatives.

Easements

Any easement relating to the site should be investigated to determine any hazard that could affect the security of the building, e.g. access to utilities under or near the building.

Planning permission

The location of the site should be such that there is unlikely to be a Local Authority requirement for the incorporation of shops or public car parking into the development, or any other requirement detrimental to security.

Urban sites

Some buildings in urban areas may have no perimeters that can be defined by a barrier. Furthermore, planning considerations may not allow the erection of a barrier higher than 1 m on road frontages or 2 m elsewhere. A fence or a wall may be impracticable and/or unacceptable to the Local Planning Authority. In city centre sites, where buildings are built up to the boundary abutting a pavement, specific hardening measures to resist blast and enhance general physical security may be required.

Where possible, buildings should be set at least 15 m from the public road. If this cannot be attained, it may be necessary to employ a form of building construction more resistant to explosive effects than would otherwise be required.

Lighting

Lighting is an important aid to physical security and a powerful deterrent to intrusion. Area lighting is used to illuminate the area between the perimeter and the building.

Integral garages

An underground car park for general use is undesirable from a security point of view. Where it is unavoidable, particular care must be taken in the planning of the accommodation immediately above and around it. There may be occasions when the separating slab has to be strengthened beyond the basic structural requirements. Sensitive facilities such as telephone exchanges should preferably not be in a basement adjacent to a garage but, if unavoidable, a solid reinforced concrete wall with no interconnecting door should separate them.

Site car parking

In most cases, car parks at street level are adjacent to the building. The objective should therefore be to ensure that no parking can take place closer than 15 m to sensitive or vulnerable parts of the building. This distance should be increased where possible. Physical barriers such as bollards prove very effective. Where site conditions prevent such an arrangement, it may be necessary to consider strengthening the part of the structure adjacent to the parking area.

Public car parking and street parking

Permanent public car parks and street car parking adjoining or overlooking the site also present a potential hazard to a building. These problems can be minimised if the Local Authority can be persuaded to restrict parking (using double yellow lines) or ban parking in the vicinity to a sufficient stand-off distance, either permanently or during a period of increased threat.

Cable routing

Normal day-to-day services such as gas, telephone, electricity cables, and pipes should be planned in a way that ensures that a prompt service is delivered to customers in the event of a disaster occurring, hence minimising the level of disruption caused. Cables entering buildings on two sides will reduce the possibility of total service disruption, as will duplication of service connections to create redundancy.

4.2 Access control

An explosive device placed inside a building causes greater damage and potentially more injuries than a similar-sized device outside a building. The installation of an access control system on both the pedestrian and vehicular entrances will therefore minimise the opportunity for the deployment of the majority of explosive devices within the building, and thus minimise the hazard to both people and property.

Where a building has vehicle and pedestrian entrances, these entrances should be separate but adjacent. A vehicle entry may need to be provided with security check points at the front and rear ends. The access road from the public highway to the vehicle entrance should provide sufficient waiting space off the public highway for vehicles awaiting entry clearance. Power-operated vehicle barriers and lifting steps may be provided in addition to gates, and should be controlled from within the building.

Loading bays and service entrances should be provided with steel roller shutters or other substantial doors, which should normally be secured from the inside only. If the entrances are to be in constant use, they should be monitored all the time.

Where vehicle access control systems are employed, they should be capable of being overridden to permit the unimpeded access of the emergency services. Care should be taken to ensure that security measures are not incompatible with fire and emergency requirements.

4.3 Bomb shelter areas

An alternative or additional preventive measure is to provide a bomb shelter area, which is well screened and protected from external attack and has sufficiently secure access control to have a very low risk of internal attack. Advice on the establishment of a bomb shelter area is presented in Appendix A.

4.4 Management procedures

The level of personnel injury and the extent of financial loss an organisation may suffer can be reduced by employing *Safety Management Practices*. These management practices should be employed by employers, landlords, or building owners and should address the protection of personnel and the minimisation of financial loss.

A full list of management measures can be obtained from *Bombs - protecting people and property*[4].

5 BUILDING DESIGN

The aim of the building design philosophy should be to minimise the consequences to the structure and its inhabitants in the event of an attack succeeding in creating an explosion. A primary requirement is the prevention of catastrophic failure of the entire structure or large portions of it. It is also necessary to minimise the effects of blast waves transmitted into the building through openings and to minimise the effects of projectiles on the inhabitants of a building.

5.1 Requirements of the Building Regulations

Control over the construction of buildings in the UK is exercised through the Building Regulations[6], a statutory document approved by Parliament under the Building Act. The aim of the Regulations is to ensure that safe buildings are constructed and that building stock is of a certain quality.

The Regulations do not contain any specific clauses that require a building to protect its inhabitants against the effects of terrorism, however they contain requirements relating to robustness and integrity of structures.

The terms *robustness* and *structural integrity* have become synonymous with the tragedy that occurred in 1968 at a 22-storey block of flats in London known as Ronan Point. A gas explosion on the 18th floor resulted in the collapse of all corner dwellings above that level and the majority of the dwellings below. It was realised that the lack of robustness and integrity of the structure, because of the lack of positive attachment between individual components, had resulted in a progressive collapse, the extent of which was disproportionate to the cause.

The guidance in the Building Regulations is contained in sections 5 and 6 of that document and deals with reducing the sensitivity of the building to disproportionate collapse. Annexes A3 and A4 respectively state that:

> "The building shall be constructed so that in the event of an accident the building will not suffer collapse to an extent disproportionate to the cause."

> "The building shall be constructed so that in the event of failure of any part of the roof, including its supports, the building will not suffer collapse to an extent disproportionate to that failure."

Other European countries have not made the same regulatory response as the UK, however this aspect is now being addressed with the development of European standards (see Section 5.2).

5.2 Requirements of the Codes of Practice

The requirements of the Building Regulations are implemented in design by Codes of Practice. British Standard Codes of Practice include:

- BS 5628: Code of Practice for use of masonry[7]
- BS 5950: Structural use of steelwork in buildings[8]
- BS 8110: Structural use of concrete[9].

Each of these design codes is based on the limit state philosophy for strength and serviceability.

In addition to the British Standards listed above, a European Standard is currently being developed that considers accidental loading such as explosions specifically. Although currently only a Prestandard, ENV 1991-2-7[10] provides a more detailed approach to design for accidental situations than current UK practice.

5.3 Guidance documents and technical manuals

Guidance documents and technical manuals are available, which may be used in conjunction with the Codes of Practice applicable in the UK. Most of these documents have military origins and provide guidance on predicting blast loads and response of structural systems. They include:

- Design of structures to resist nuclear weapons effects, Manual 42[11]
- Structures to resist the effects of accidental explosions: TM5-1300[12]
- A manual for the prediction of blast and fragment loadings on structures: DOE/TIC-11268[13]
- The design and analysis of hardened structures to conventional weapons effects[14]
- Structural design for physical security[15].

Each of these documents is described briefly below.

Design of structures to resist nuclear weapons effects, Manual 42

This document provides guidance to engineers engaged in designing facilities to resist the effects of nuclear weapons. Although emphasis is placed on blast-resistant design, other effects are treated in some detail to give the reader sufficient background to seek additional information in the published literature. The document contains information that the engineer can use in developing design criteria and in establishing analysis and design procedures.

Structures to resist the effects of accidental explosions: TM5-1300

This manual is regarded as a principal source, and is the one most widely used by both civilian and military organisations for designing structures to prevent the propagation of explosion and to provide protection for personnel and valuable equipment. Step-by-step analysis and design procedures are presented in the manual and detailed information is given on:

- blast, fragment, and shock loading
- principles of dynamic analysis
- reinforced concrete and structural steel design
- special design considerations.

Guidance is also provided for selection and design of security windows, doors, utility openings, and other components that must resist blast and forced entry effects. The manual contains a valuable listing of most relevant references.

Distribution of the document is unrestricted.

A manual for the prediction of blast and fragment loadings on structures DOE/TIC-11268

This manual provides guidance to the designers of facilities subject to accidental explosions and aids the assessment of the explosion-resistant capabilities of existing structures. The document is used in conjunction with other structural design manuals.

Distribution of the document is unrestricted.

The design and analysis of hardened structures to conventional weapons effects

This manual is based on state-of-the-art design information and methods for protective structures and includes new, recently analysed and validated test data from the Defense Nuclear Agency test programmes on conventional weapons effects, together with design examples. The document supersedes TM 5-855-1 *Fundamentals of protective design for conventional weapons*[16] and ESL-TR-87-57 *Protective construction design manual*[17].

The document is supported by interactive software that contains text, figures, graphs, tables, equations, and a number of stand-alone computer codes such as BLASTX and FOIL (see Section 7.8), which are written for DOS- and UNIX-based operating systems.

Certain sections have unrestricted distribution, but the manual as a whole has restricted distribution.

Structural design for physical security

This is a comprehensive guide for civilian designers and planners wishing to incorporate physical security considerations into new designs or building refurbishment projects.

5.4 Choice of structural system

The construction materials, structural form, and connections used are all factors influencing the ability of a structure to withstand blast loads, however certain types of construction commonly used in *ordinary buildings* are not generally suitable for blast-enhanced structures. Blast-protected buildings should generally be designed and detailed to absorb the blast energy by controlled deformation.

Hence the principle basis of evaluating their suitability is an assessment of their mode of failure under severe overload.

Brittle construction of the primary structure of the building is not suitable: besides being vulnerable to catastrophic sudden failure under blast load, it provides a source of debris that can cause major damage when hurled by the blast wind. Non-reinforced concrete, brick, timber, masonry, and corrugated plastic are examples of brittle construction, however these forms may be specified in the exterior shells of blast-protected structures if venting of blast energy is required to reduce blast overpressures. If, in an otherwise ductile structure, brittle behaviour of some elements cannot be avoided, the margin of safety for these elements should be increased, i.e. their capacity should be downgraded.

Framed buildings built from a steel or reinforced concrete skeleton are generally suitable if the connections provide full continuity. The desirable protective features found in framed construction are summarised by Christopherson[18] as:

- Ductility - individual members or sections of the structure can be subjected to very large distortions before they become incapable of carrying load.

- Redundancy - even when a limited number of members can no longer carry load, the structure as a whole will be capable of redistributing the load among the remaining members and thus avoiding total collapse. Many practical frames are in fact highly redundant (although not designed with that in mind) and could be made even more so by small changes in connections.

- Permeability - the blast load is able to travel freely through the spaces between members, which present only a relatively small area to a wave travelling in any direction. The impulse on the frame will be reduced.

Steel-framed buildings have joints between beams and columns that are either simple shear or continuous connections. Simple shear connections can resist blast forces adequately providing they are detailed such that they can distort yet continue to function as load-carrying members. The advantages with continuous joints are that they offer more load paths and more opportunity for developing plastic hinges with sufficient rotation to absorb energy. Hence, if a member is removed as a result of a blast load, forces in members and joints will be redistributed along alternative paths, in the manner of a Vierendeel lattice. Once again, correct detailing of joints is important to ensure that full-strength joints are obtained.

Under blast loading, buildings are subjected to loads that are quite different from those governing their primary design in both magnitude and direction. For example, structures are naturally strong under downwards vertical loading but are potentially vulnerable against either high upwards loading on floors or high lateral loading. To avoid disproportionate collapse, it is important that these alternative loading modes are examined. From experience, a potentially weak zone is the connection of the external columns to the internal floors. High internal pressure generates a tensile force absent under normal loading conditions, and inadequate strength allowing the external columns to be detached may precipitate significant damage.

Diaphragm sway walls greatly increase the protection that the building provides to terrorist attack. They also provide excellent protection against projectiles and localise the effect of primary blast. Furthermore, because there is some latitude as regards where in a framework diaphragm walls are located, they can be positioned to gain the greatest protective advantages.

The rules for effective blast resistance can conflict with each other when several loading cases are considered. This is particularly so when countering terrorism, as there can be a diversity of threat. In particular, difficulties arise where characteristics have both good and bad effects depending on the direction of loading. For example, openings allow venting from low-order internal explosions to provide *permeability*, thereby reducing blast impulse, however openings greatly increase the threat from the most likely cause of injury following an explosion, i.e. secondary missiles (such as flying glass). One solution to allow venting but prevent secondary missiles from external explosions is to use blast-relieving panels.

Generally speaking, steel-framed and reinforced concrete-framed buildings are both good at withstanding explosions, however it is much easier to make steel-framed structures ductile by design and detailing than it is for reinforced concrete frames. Although both types of framed systems are capable of absorbing large amounts of energy through plastic deformation, ductility will only be achieved if care is taken in detailing. For steel frames, this means that joints are made full-strength connections and sections are sized such that member instability, preventing full rotation at plastic hinges, does not occur. For reinforced concrete frames, ductility will only be present if additional reinforcement is included that prevents spalling of the concrete surface and does not allow the concrete core to disintegrate. Even if individual members fail, significant collapse will not occur as alternative load paths are available. One benefit of choosing steel frames is that enhancement or repair of existing steel members can easily be undertaken. For reinforced concrete-framed buildings, more radical modifications may be required, which can make repair difficult and/or expensive. Rhodes[19] found that the most common point of failure was inside reinforced concrete joints, where the concrete was reduced to rubble. Although collapse had not occurred, this damage was difficult or impossible to reinstate and part of the building had to be replaced. As noted above, these situations can be avoided if appropriate detailing is adopted in the original design.

6 EXPLOSIVES

This Section describes how an explosion caused by a high explosive source occurs. The various types of high explosives are described together with the devices and methods of delivery that have been used. As there are many forms of high explosive, a method is described that enables the explosive effect of a particular explosive to be determined using the TNT explosive as a basis.

6.1 Sources of explosion

Explosions can be caused by the following:

- chemical reactions (e.g. high explosives)
- thermonuclear reactions
- the ignition of combustible products
- dust cloud explosions.

Only explosions caused by high explosives (chemical reactions) are considered in this document, although some reference is also made to incendiary devices.

The explosion characteristics from the other sources are completely different from those of high explosives, and protection against them requires very different considerations.

6.2 Explosive materials

High explosives are solid in form and are commonly termed condensed explosives. These explosives can be classed into the following:

- military or commercial explosive
- home-made explosive
- home made incendiaries.

Military or commercial explosives (M/CE) contain carbon, hydrogen, oxygen, and nitrogen (CHON) compounds; TNT (trinitrotoluene) is perhaps the most widely known example. In military use, TNT is used as the filler for munitions where it is usually mixed with other high explosives such as cyclonite (RDX), HMX (a complex CHON compound), or ammonium nitrate. For commercial use, explosives tend to be mixed with cheaper nitrates. Semtex is one form of commercial high explosive that has been used by terrorist organisations, albeit in limited quantities.

Home-made explosives (HME), as the name implies, are explosives manufactured using readily-available chemicals such as farm fertilisers (ammonium nitrates). The fertiliser can be mixed with commercial explosives to produce enhanced composition. Because of their variability, it is difficult to define a home-made explosive accurately.

Home-made incendiaries contain accelerants, and are commonly liquid fuels such as petrol or white spirits. In certain cases, high explosives are added to distribute the accelerants so that damage is more widespread.

6.3 Explosion process for high explosives

The violent release of energy associated with the detonation converts the explosive material into very hot gases (typically at 3000°C), which rapidly expand radially away from the source of the detonation. As these gases are partially restrained by the surrounding air, the pressures generated within the gases can be as high as 300 kilobar (3×10^7 kPa). Because of the restraining effect of the surrounding air, a layer of compressed air is formed. This layer is called a shock front or, more commonly, a blast wave. This blast wave travels radially away from the point of detonation with a diminishing shock velocity, although it is always in excess of the velocity of sound in air. As the shock front travels away from the explosion source and expands into an increasingly larger space, its strength decays.

Gas molecules behind the shock front that move at a lower velocity than that of the shock front cause dynamic pressures to be generated. These dynamic pressures are caused by the wind generated by the passage of the shock front.

As the pressure of the explosive gases reduces, the gases expand. Eventually their pressure reduces to a little below ambient, resulting in an *over-expanded* state. As a consequence, a pressure differential is generated between the combustion gases and the atmosphere, causing a reversal in the direction of flow, back towards the centre of the explosion, known as a *negative pressure phase* or *suction phase*. This is a negative pressure relative to atmospheric, rather than absolute negative pressure. Equilibrium is reached when the air has returned to its original state.

The duration of high-explosive explosions is commonly of the order of tens of milliseconds. This contrasts with fuel-, gas-, or dust-based explosions, where the duration is typically hundreds of milliseconds.

6.4 Explosive devices and methods of delivery

There are many ways in which an explosive device may deliver an attack; the method used at a particular time will depend on many factors. The type of device used will reflect the terrorists' objectives and access to explosive material, and will also be influenced by the defensive measures in place (e.g. bag searches will reduce the threat of briefcase bombs).

In the UK, several trends have emerged with regard to the type of device used, and these reflect the past political climate. The more common devices, sizes, and methods of delivery are summarised in Table 6.1. This list is not exhaustive and should only be used as guide based on past activities.

Table 6.1 *Examples of typical terrorist devices used in Western Europe*

Category	Method of delivery	Typical explosive material*	Typical charge weight used (approx.)	Possible damaging effects
Culvert bombs	Pre-planted	HME	Unlimited	Blast
Mortar bombs	Mortar tubes	M/CE	≤ 200 kg	Blast + impact momentum
	Letter	M/CE	50-40 g	Blast/fire
	Hand thrown	M/CE	2 kg	Blast
Package bombs	Small briefcase	M/CE	2-4 kg	Blast + fragment
	Large briefcase	M/CE	4-12 kg	Blast + fragment
	Suitcase	M/CE	12-22 kg	Blast + fragment
	Bike	M/CE	30 kg	Blast
	Under vehicle	M/CE	0.5-1 kg	Blast
Vehicle bombs	Car	HME	≤ 250 kg	Blast/fragment + impact
	Small van	HME	250-900 kg	Blast/fragment + impact
	Large van	HME	1-2 t	Blast/fragment + impact
	Small lorry	HME	2-3 t	Blast/fragment + impact
	Medium lorry	HME	3-4 t	Blast/fragment + impact
	Large lorry	HME	4-5 t	Blast/fragment + impact
Incendiary devices	Standard	HMIC	30-120 g	Fire
	Blast	M/CE + HMIC	30-120 g	Blast/fire

* HME   Home-made explosive, e.g. small quantity of M/CE + ammonium nitrate.
  M/CE   Military or commercial explosive, e.g. trinitrotoluene (TNT).
  HMIC   Home-made incendiary composition, e.g. petrol, white spirits.

6.4.1 Vehicle bombs

Vehicle bombs are bombs contained in cars, vans, trucks, or any other form of motorised transport. These bombs usually contain large amounts of home-made explosives, complemented by a small quantity of military or commercial explosive to ensure detonation. The vehicles containing the explosives may either be driven at speed into a target or they may be left to explode.

6.4.2 Package bombs

Hand-carried bombs are normally classed as package bombs. Package bombs can range from letter bombs weighing a few grams to suitcase-sized bombs weighing 10-20 kg. They can be placed either outside a target (with reduced risk of being detected) or inside the target building, either by the terrorists themselves or by proxy. Package bombs can be disguised in many containers, e.g. bins, bicycles, thermos flasks, and letters.

Package bombs have been placed under vehicles, in such a way that the user of the vehicle is the target. The explosive material normally used for such an attack is a small quantity of military or commercial explosive.

7 BLAST OVERPRESSURE AND LOADS

This Section introduces the different types of explosion that can occur and explains how the blast-wave characteristics caused by the explosion can be obtained and used to determine blast loads on buildings.

7.1 Types of explosion

Explosions caused by high explosives can be:

- unconfined explosions
- confined explosions
- explosive attached to a structure.

7.1.1 Unconfined explosions

Unconfined explosions can be subdivided into:

- free-air burst
- air burst
- surface burst.

Free-air burst

A free-air burst explosion is one where the explosion occurs in free air high above ground level (Figure 7.1). The spherical shock wave caused by the detonation of the high explosive strikes a building without any intermediate amplification of the blast wave. Reflected waves are created when the front strikes the building (see Section 7.4).

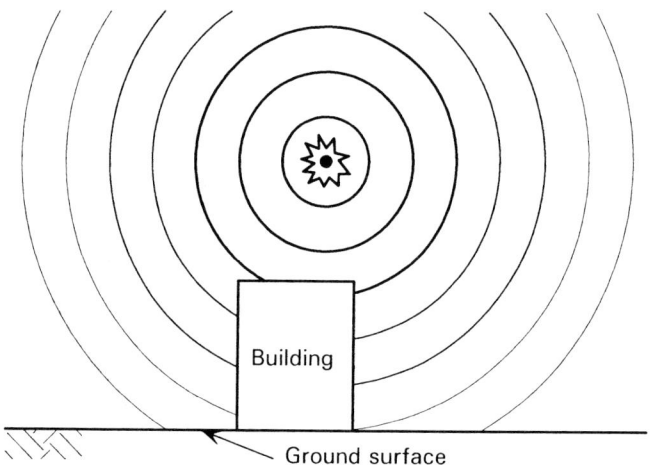

Figure 7.1 *A free-air burst explosion*

Air burst

In an air-burst explosion, the detonation of the high explosive also occurs above ground level, but in this case intermediate amplification of the wave caused by ground reflections occurs prior to the arrival of the initial blast wave at a building (Figure 7.2).

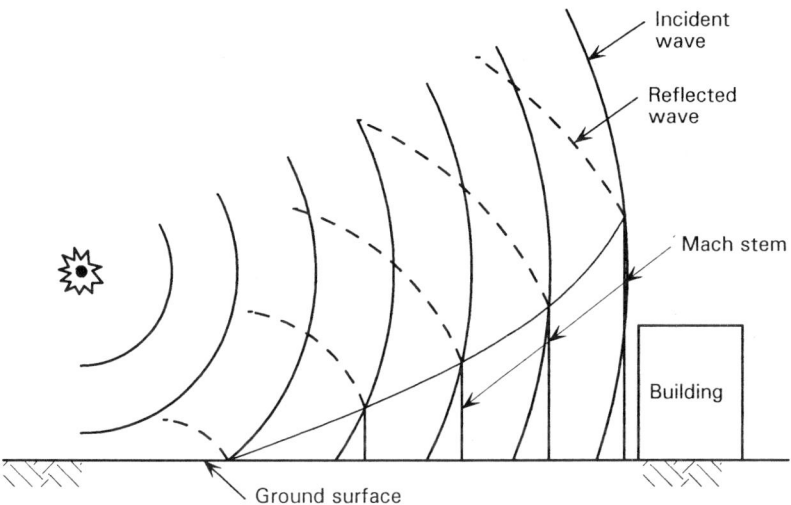

Figure 7.2 *An air burst with ground reflections*

As the shock wave continues to propagate outwards along the ground surface, a front commonly called a Mach stem is formed by the interaction of the initial wave and the reflected wave. The characteristics of the Mach stem are described in Section 7.4.3.

Surface burst

A surface-burst explosion occurs when the detonation occurs close to or on the ground surface. The initial shock wave is reflected and amplified by the ground surface to produce a reflected wave (Figure 7.3).

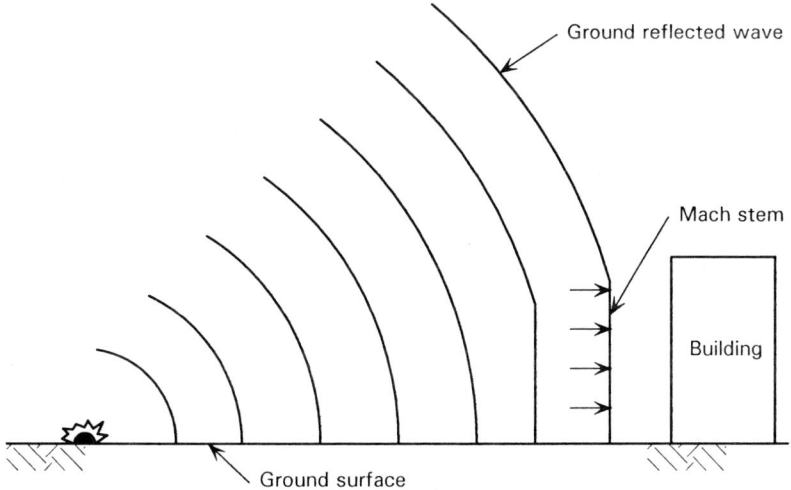

Figure 7.3 *Surface burst*

Unlike the air burst, the reflected wave merges with the incident wave at the point of detonation to form a single wave, similar in nature to the Mach stem of the air burst. Essentially, a hemispherical shock front is created (Figure 7.4). The characteristics of the Mach stem are described in Section 7.4.3.

Figure 7.4 *An explosion caused by a surface burst [Courtesy of TPS Consult]*

In the majority of cases, terrorist activity has occurred in built-up areas of cities, where devices are placed on or very near the ground surface. Because of the closeness of tall and high-rise buildings, it is necessary to consider the impingement of surface bursts and blast-wave reflections (single or multiple) when determining blast loading acting on buildings.

7.1.2 Confined explosions

When an explosion occurs within a building, the pressures associated with the initial shock front will be high and will in turn be amplified by their reflections within the building. In addition, and depending upon the degree of confinement, the effects of the high temperatures and accumulation of gaseous products produced by the chemical reaction involved in the explosion will exert additional pressures and increase the load duration within the structure. Depending on the extent of venting, various types of confined explosions are possible. These explosions can be classed as:

- fully vented
- partially vented
- fully confined.

The three classes are shown schematically in Figure 7.5.

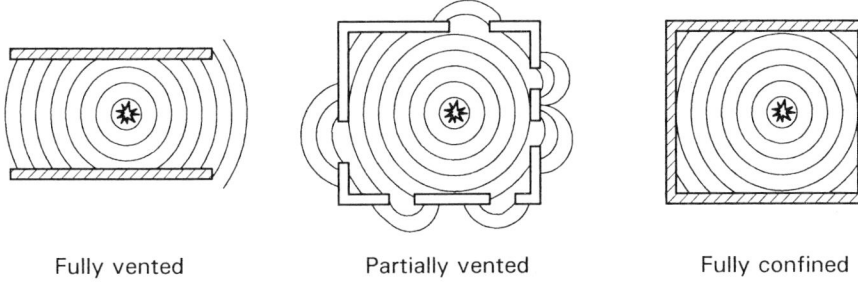

Fully vented   Partially vented   Fully confined

Figure 7.5 *Fully-vented, partially-vented, and fully-confined explosions*

Fully-vented confined explosions

Fully-vented explosions are produced within a building where one or more of the surfaces is open to the atmosphere. The initial blast wave is totally vented almost instantaneously to atmosphere, forming a blast wave that propagates away from the confinement.

Partially-vented confined explosions

Partially-vented explosions are produced in a building where the extent of open surfaces is more limited than in the fully-vented case. The initial blast wave is vented to atmosphere after a finite period of time.

Fully-confined explosion

A fully-confined explosion is produced when there is either total or near total containment of the explosion by the surfaces forming the barrier or building.

7.1.3 Explosive in contact with a structure

If a detonating explosive is in contact with a structural component (e.g. a column or beam), the arrival of the detonation wave at the surface of the explosive will generate intense stress waves in the material, producing crushing, shattering, or disintegration of the material. This hammer-blow effect is called *brisance*.

An explosive in contact with a structure also produces the effects described in Sections 7.1.1 and 7.1.2.

7.2 Scaled distance

The fundamental parameter that is used extensively to determine blast-wave characteristics is called the *scaled distance*. The principle of the scaling law is based on conservation of momentum and geometric similarity. The relationship, formulated independently by Hopkinson[22] and Cranz[23], is described as cube-root scaling and is defined as:

$$Z = \frac{R}{W^{1/3}}$$

where Z is the scaled distance, with units m/kg$^{-1/3}$
             R is the range from the centre of the charge
             W is the mass of spherical TNT charge (see Section 6.5).

The use of scaled distance Z allows a compact and efficient presentation of blast-wave characteristics for a wide range of situations (see Sections 7.3 and 7.4).

7.3 Blast-wave characteristics in free air

7.3.1 Blast-wave overpressure profile

The progression of a free-air burst is best represented by a pressure-time history curve. The magnitude of the pressure caused by a blast wave is usually quoted as an overpressure, i.e. the pressure increase relative to ambient pressure. The characteristics of the blast wave that are commonly used for calculation purposes are shown on a blast overpressure-time curve presented in Figure 7.6. The curve shows the variation of blast overpressure with time at a particular location within the blast-affected region.

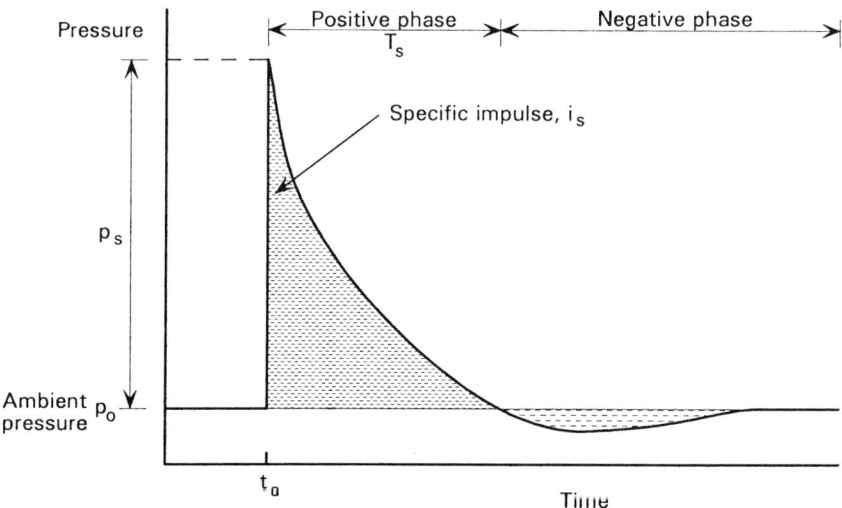

Figure 7.6 *Blast overpressure-time profile*

The arrival of the steep shock front is assumed to be an essentially instantaneous increase of pressure. The shock front is followed by a fairly rapid decay in positive overpressure (pressure greater than ambient pressure). This positive overpressure is followed by a period of negative overpressure (pressure below ambient pressure) until sustained normal ambient pressure returns.

7.3.2 Parameters describing the blast-wave characteristics

The parameters that describe the characteristics of a blast wave are defined for free-air burst conditions and are commonly called *side-on* blast-wave parameters. The most commonly used parameters include:

p_s the peak side-on overpressure for positive overpressures
T_s the duration of the positive overpressure phase
i_s the specific side-on impulse associated with the positive overpressure phase
t_a the arrival time of the blast front.

These parameters can be determined most readily from empirical relationships obtained from experimentation and are presented in numerous publications, including TM5-1300[12] and *Explosion hazards and evaluation*[20]. The parameters are best presented graphically as a function of the scaled distance parameter Z (see Figure B.1 in Appendix B).

Various additional parameters can be determined that characterise the negative overpressure region of the blast overpressure profile (e.g. the peak underpressure magnitude and underpressure duration), however these are not usually required for the simplistic dynamic response analyses.

It is important to be aware that close to the charge (near field), i.e. for $Z < 0.5$ m/kg$^{1/3}$, a noticeable spread of results occurs, so the accuracy of predictions will be lower.

7.3.3 Dynamic pressure

After the initial blast wave or shock front has passed, a blast-induced wind, which consists of air, gases, and combustion products, causes dynamic pressures to be generated. These dynamic pressures cause drag loads to be experienced by the building. Standard relationships (in a free field situation) plotted against scaled distance have been established between the peak dynamic pressure q_s, the gas particle velocity $\bar{u}_s$ ($= u_s/a_o$), and the velocity of the shock front $\bar{U}$ ($= U_s/a_o$) (a_o is the velocity of sound in ambient conditions). These are presented as graphical plots in Figure B.2 in Appendix B.

Table 7.1 shows peak dynamic pressure q_s for a selection of peak side-on overpressures p_s. Dynamic pressures are not small and can be greater than peak side-on overpressures. It is assumed that dynamic pressures act for the same duration as blast-wave overpressures.

Table 7.1 *Comparison of dynamic pressures with peak side-on overpressures*

p_s (kPa)	q_s (kPa)
100	31
300	225
500	518
700	875

7.4 Reflected blast-wave characteristics

A blast wave that impinges on a solid surface of a building (or on a dense medium) will be reflected. These reflections, particularly in built-up areas, create complex loading conditions and in certain circumstances can lead to a succession of high magnitude shock fronts that are amplified by the reflection process.

Three types of reflection phenomenon can occur:

- face-on
- regular
- Mach.

These depend on:

- the incident angle of the blast wave on the reflecting surface
- the incident overpressure of the blast wave
- the ambient pressure of the air
- the type of reflecting surface (i.e. its ability to absorb energy).

For determination of blast pressure caused by unconfined explosions, it is common to make a conservative assumption that the surface of the building or structure is unyielding, i.e. none of the blast-wave energy is lost due to the deformation of the surface.

Each type of reflection produces significantly different blast loading characteristics.

7.4.1 Face-on reflection

Face-on reflection occurs when a blast wave strikes a surface oriented at right angles to the direction of the wave front (Figure 7.7).

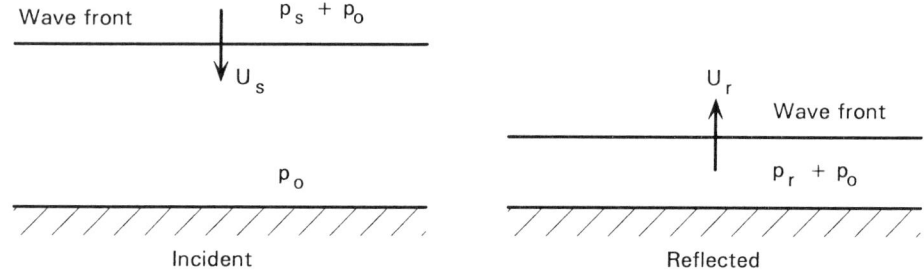

Figure 7.7 *Normal or face-on reflection of a blast wave*

The air molecules at the front of the blast wave are stopped abruptly by the presence of the unyielding surface. These molecules are compressed by the trailing pressure wave, which causes the reflected overpressure to be greater in magnitude than the incident pressure. The pressure-time profile for a reflected wave is shown in Figure 7.8.

It can be seen that the increase in blast overpressure causes a corresponding increase in the impulse of the blast wave. Time durations on the other hand are assumed to remain unaltered and are equal to those given by free-air characteristics.

Face-on reflection blast-wave parameters can be obtained from Figure B.3 in Appendix B, for the value $\alpha = 0$.

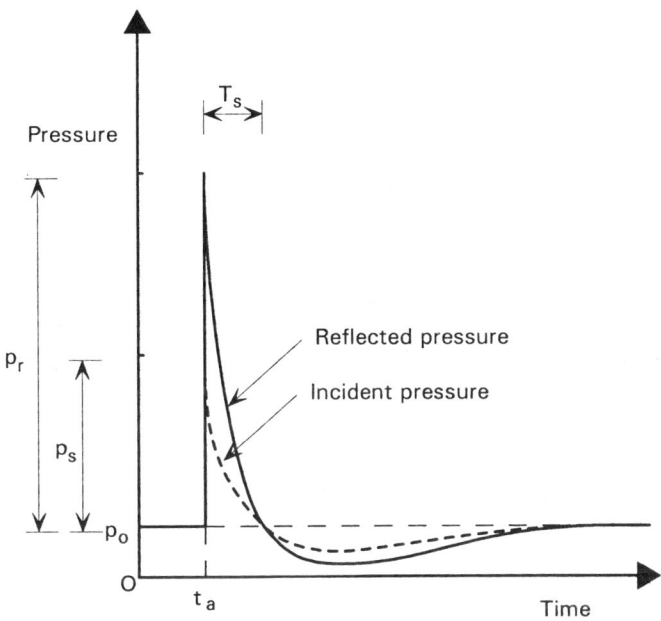

Figure 7.8 *Blast-wave overpressure profile for a face-on reflected wave*

7.4.2 Regular reflection

Where the incident angle of the blast wave α_i lies between $0°$ (face-on incidence) and approximately $40°$ (in air), regular reflection occurs (Figure 7.9).

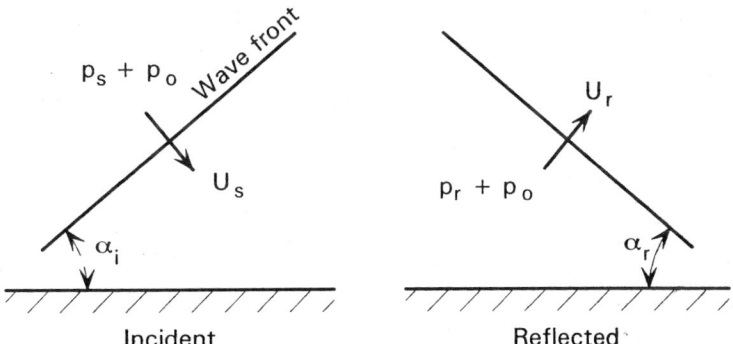

Figure 7.9 *Regular reflection of a blast wave*

The reflected pressure caused by regular reflection is greater than the reflected pressure caused by face-on reflection. Also, the reflected angle will not equal the incident angle, but as the incident angle increases so does the reflected angle.

The peak overpressure of the reflected wave front can be obtained from Figure B.4 in Appendix B. The reflection coefficient C_r is given by

$$C_r = \frac{\text{peak reflected overpressure}}{\text{peak incident overpressure}} = \frac{p_r}{p_s}$$

Peak reflected overpressure depends on the magnitude of the peak incident overpressure and for values of α_i up to $40°$ lies between two and 13 times the peak incident overpressure (see Figure B.4 in Appendix B).

7.4.3 Mach reflection

When the critical or limiting incident angle for a given reflected pressure is exceeded, regular reflection is replaced by Mach reflection. In air, this limiting angle is approximately 40° (Figure 7.10).

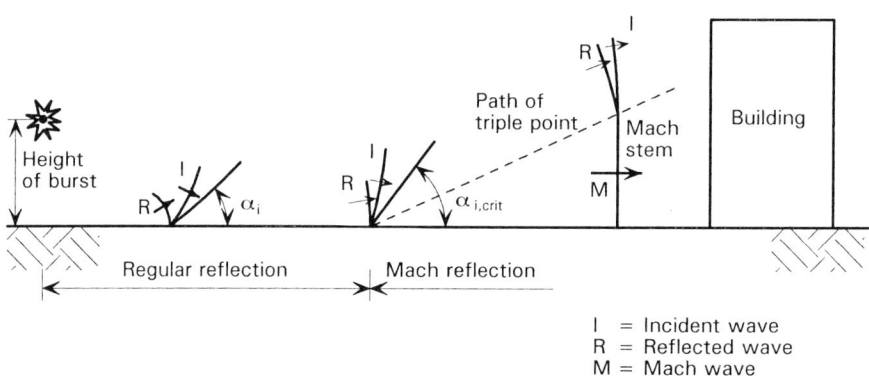

Figure 7.10 *Development of Mach stem and Mach reflection*

The mechanism of Mach reflection is a result of the reflected wave catching up with the incident wave. The reflected wave cannot overtake the incident wave, so the two wave fronts join at some distance away from the reflecting surface. The wave front produced is called the Mach stem. The formation of a Mach stem is shown in Figure 7.10. The height of the Mach stem increases as the wave propagates away from the centre of detonation and is assumed to be a plane wave over its full height. The triple point is formed at the intersection of the incident, reflected, and Mach waves.

Mach reflection can cause a substantial increase in reflected overpressure, and can lead to much higher overpressures than predicted by assuming that the blast wave undergoes regular reflection. The extent of magnification depends on the initial incident peak overpressure magnitude.

Mach stems are particularly important in internal and city centre explosions where multiple and shallow angles of reflection are common, however the magnitude of the Mach stem reflected overpressure should not be taken out of context. In the case of an explosion at the base of a multi-storey building, the Mach stem is formed at a height where the natural decay of the incident overpressure is such that the magnitude of the Mach stem is not very significant. In practice, therefore, the regular reflected overpressures will be similar in magnitude to the Mach stem effects (Figure 7.11).

The reflection coefficient C_r for Mach reflection can be obtained from Figure B.4 in Appendix B.

Further information on the behaviour of reflected blast waves is given by Baker *et al.*[20], Bangesh[21], and Kinney and Graham[24].

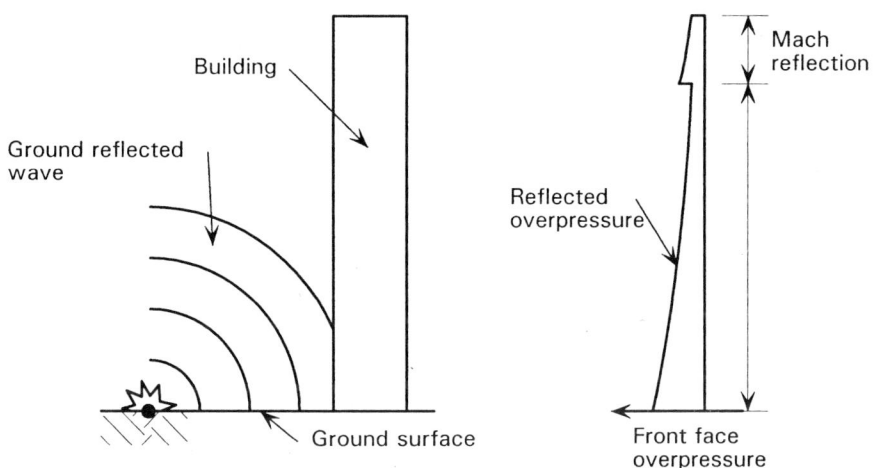

Figure 7.11 *Overpressure versus distance in the regular reflection and Mach reflection regions*

7.5 Blast overpressures due to a surface burst

In the preceding Section, incident blast-wave characteristics have been presented for free-air explosions that are assumed to be spherical air bursts remote from any reflecting surface. For a surface burst, however, the blast wave front is not spherical but hemispherical, as discussed in Section 7.1.1 and shown in Figures 7.3 and 7.4.

Based on the correlation of experimental data between spherical free-air bursts and hemispherical surface bursts, the characteristics of a blast wave due to surface burst can be determined accurately from free-air characteristics by assuming that the mass of the explosive W is modified by an enhancement factor, which is normally taken to be 1.8. This enhancement factor is applied before the incident blast-wave parameters are calculated.

In theory, if the ground surface were a perfect reflector, the enhancement factor would be equal to 2, but as energy generated by the explosion is partially absorbed by the ground, resulting in the formation of a crater and ground shock, the enhancement factor has to be reduced. The relationship can be expressed approximately as:

$$W_{\text{surface burst}} = 1.8W$$

where $W_{\text{surface burst}}$ is the enhanced mass of explosive when placed on or very near the ground surface
W is the TNT equivalent mass of explosive.

7.6 Mathematical representation of blast overpressure profile

The blast overpressure profile $p(t)$ for positive blast overpressures can be represented mathematically by the following relationship (illustrated in Figure 7.6):

$$p(t) = p_s \left(1 - \frac{t}{T_s}\right) \exp\left(-\frac{bt}{T_s}\right)$$

where p_s is the peak side-on overpressure
  T_s is the duration of the positive phase of the blast
  b is the waveform parameter
  t is time measured from the instant that the blast wave arrives (at time = t_a).

Values of these parameters are obtained from Figures B.1 and B.2 in Appendix B.

It is usually adequate to assume that the decay (and growth) of blast overpressure is linear. For the positive overpressure phase, a simplification is made where the impulse of the positive phase of the blast is preserved and the decay of overpressure is assumed to be linear. This simplification is shown in Figure 7.12.

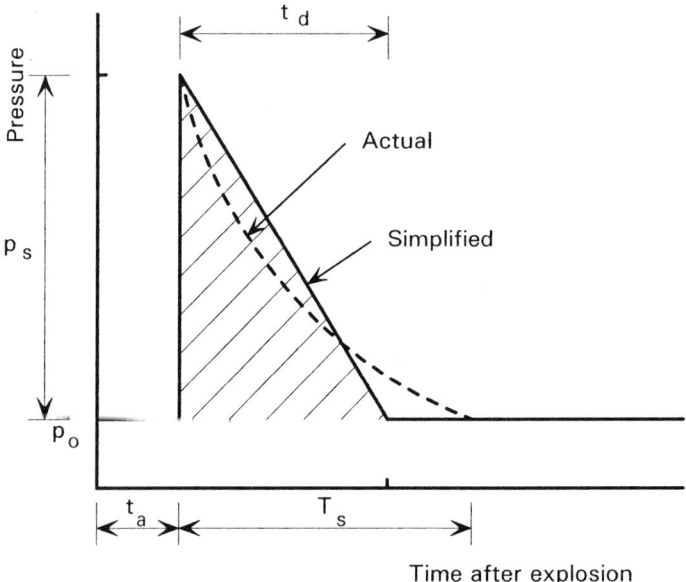

Figure 7.12 *Simplification of the blast-wave overpressure profile with impulse preserved*

For this simplified case, the blast overpressure-time relationship is given by

$$p(t) = p_s \left(1 - \frac{t}{t_d}\right)$$

where t_d is the positive phase duration of the simplified linear blast-wave overpressure profile.

This equation is considered adequate in predicting blast pressures because the positive overpressure phase is most significant in determining structural response when using the simplistic models described in Section 8. If required, however, similar simplifications can be made to the negative overpressure-time profile and a revised relationship developed. Peak negative overpressures, duration, and impulse can be obtained from Baker *et al.*[20]

7.7 Blast pressures and loads on buildings

In reality, a building is of a finite size and has a roof, sides, and a rear face, i.e. it is a solid object. Because of its finite size and solid shape, a building will experience a time-varying system of pressures that are more complex than the simple side-on and reflected overpressures presented in the previous Sections, where it was assumed that surfaces were infinite in size, and where diffraction could not occur.

The loading on exposed faces of buildings is a function of the reflected and side-on overpressures and the dynamic (blast wind) pressures; the size, shape, and orientation of the walls of the building; and the location and orientation of other objects nearby.

The following two Sections describe how loads caused by an external explosion are generated on external surfaces of a building. The effects of a large-scale explosion and a small charge explosion are described.

7.7.1 Large-scale explosion

The impingement of a large-scale explosion that engulfs a rectangular, closed building is a difficult process to explain because a full understanding requires familiarity with computation fluid mechanics. A simplistic treatise is given here without resorting to complexities. Each face of the building is considered in turn and simplified overpressure profiles are presented with accompanying simplistic descriptions of the effects including diffraction that are taking place. The publication *Explosive shocks in air*[24] gives a more thorough explanation.

The location of the building is assumed to be outside the region of regular reflection and in the region of Mach reflection. The passage of the wave front over it and the corresponding diffraction phenomena are shown in Figure 7.13.

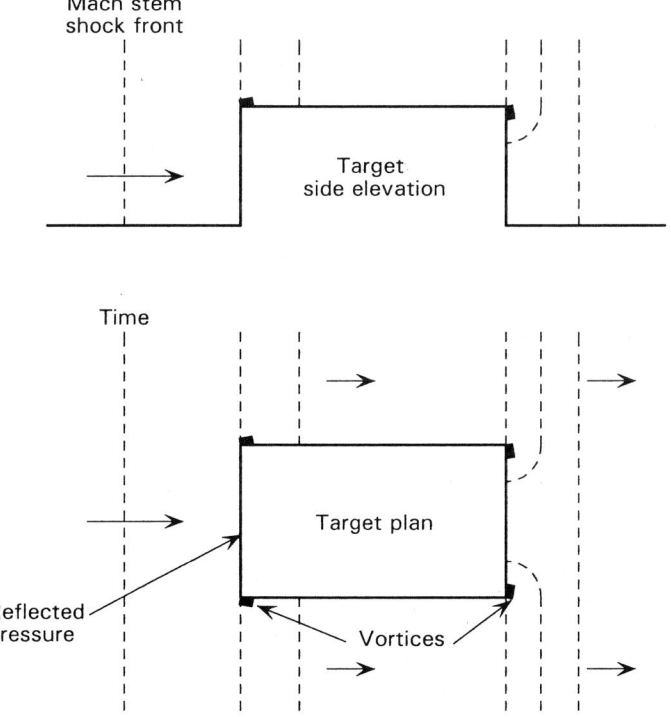

Figure 7.13 *Progression and behaviour of blast wave*[25]

Front face

The front face of the building experiences enhanced peak overpressures due to reflection of the incident blast wave. Once the initial blast wave has passed the reflected surface of the building, the peak overpressure decays to zero at a rate dependent on the duration of the blast wave (see Figure 7.14).

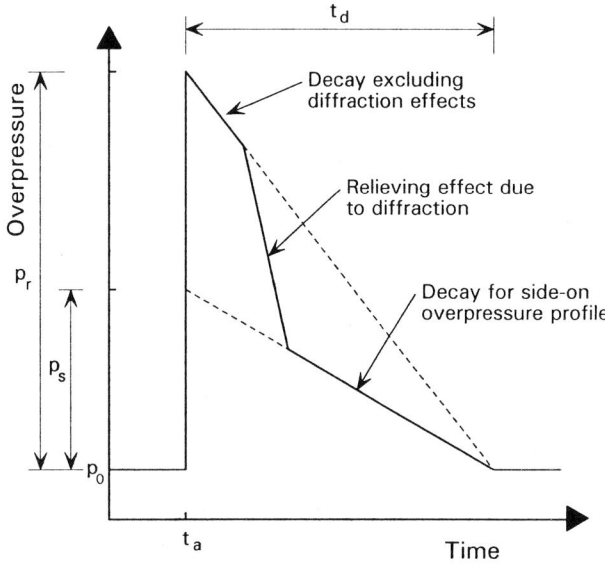

Figure 7.14 *Blast overpressure profile acting on an area of the front face of the building*

As the sides and top faces of the building in turn receive *side-on* overpressures (which are lower than the reflected overpressures on the front face), a relieving effect of blast overpressure is experienced on the front face, which causes the rate of decay of overpressure on the front face to increase. Once this relieving effect is complete, the overpressure decays at a rate similar to that experienced for side-on overpressure decay.

The modified blast overpressure profile shown in Figure 7.14 can be simplified further if required to an equivalent triangular pulse profile similar to the profile shown in Figure 7.12, however it is important that impulse for the simplified profile remains the same as that for the profile developed in Figure 7.14.

Dynamic pressures (blast wind) also act on the front face after the blast wave has passed the front face. A simplistic representation of the variation of dynamic pressure with time acting on an area of the front face of a building is shown in Figure 7.15.

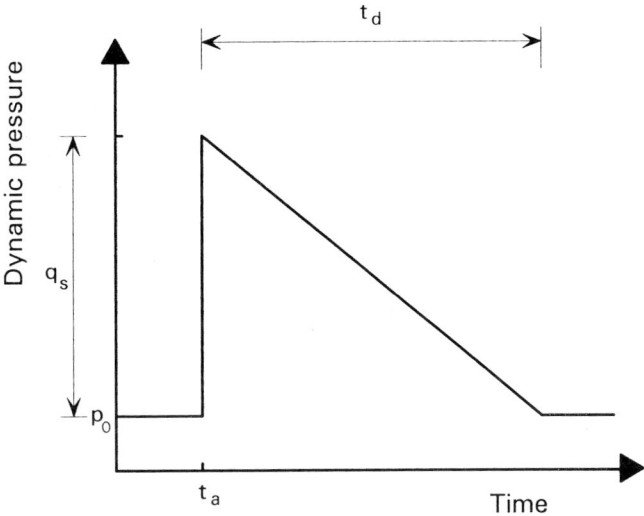

Figure 7.15 *Dynamic pressure profile acting on an area of the front face of the building*

It is assumed that dynamic pressures act for the same duration (t_d) as that for side-on and reflected blast overpressures.

Overall forces acting on the faces of the building are calculated by summing with respect to time the forces determined for each component area investigated (see Section 7.7.3).

Roof and sides

The sides and top faces of the building are exposed to *side-on* overpressure (which is the overpressure of the propagating wave in a free field environment with no reflections) (Figure 7.12). If the side walls and roofs are oriented parallel to the direction of the blast wave, the effects of drag loads caused by the blast wind are negligible.

Rear face

The rear face experiences load characteristics similar to those experienced by the front face. The rear of the structure experiences no overpressure until the blast wave has travelled the length of the structure and a compression wave has begun to move inwards, towards the centre of the rear face. In this instance, the pressure build up is not instantaneous. A typical pressure-time profile is shown in Figure 7.16.

Translational forces acting on the building

As the building has a finite depth, there will be a time lag in the development of pressures and loads on the front and back faces. This time lag causes translational forces to act on the building in the direction of the impinging blast wave. These translational forces are complex and depend on the dimensions of the building.

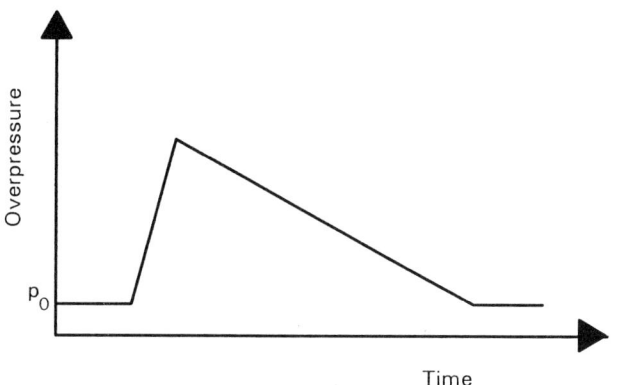

Figure 7.16 *Back-wall overpressure versus time*

7.7.2 Small-scale explosion

When a relatively small charge of short duration is detonated near a substantially-sized building, blast loading is considered to act locally on the front face of the building only, hence the front face can be assumed to be infinite. In this case, individual elements of the building can be analysed separately because the components are likely to be loaded sequentially.

The assumed blast overpressure profile will be as shown in Figure 7.12.

7.7.3 Blast load on individual surfaces

The total load that acts on an area or element of a structure consists of two time-dependent components of load, which can be represented by the relationship:

$$F(t) = F_{Blast}(t) + F_{Dynamic}(t)$$

F_{Blast} is the component of loading that is caused by the blast wave, while the second component $F_{Dynamic}$ is the load caused by dynamic pressure as a result of the airflow following the shock wave (blast wind).

The load components F_{Blast} and $F_{Dynamic}$ are given by

$$F_{Blast} = p A_{proj}$$

for the shock component and

$$F_{Dynamic} = q_s C_D A_{proj}$$

for the dynamic component

where p is the overpressure caused by the blast wave
 A_{proj} is the area of structure under consideration projected onto the plane normal to the direction of approach flow
 q_s is the dynamic pressure of the gas particles behind the blast wave
 C_D is the dynamic or drag coefficient.

The magnitude of the drag coefficient C_D depends on the geometry of the object and on the particle velocity. For the front face of the building, C_D should be taken as +1.0. For roofs, back walls, and side walls of buildings, TM5-1300[12] recommends the values dependent on pressure, as presented in Table 7.2.

Table 7.2 *Drag coefficient C_D to be used for buildings*

Peak dynamic pressure (kN/m^2)	Drag coefficient (C_D)
0-172	-0.4
172-345	-0.3
345-896	-0.2

For the back wall, the negative values of C_D indicate that suction occurs.

For a more comprehensive range of building geometries, values of C_D can be obtained from BS 6399[26].

The loading function is generally too complex to be defined explicitly, and can only be determined with some degree of accuracy if computational fluid dynamic analyses are performed (see Section 7.9). Crude approximations can be obtained using simplistic hand calculation methods for approximating forces acting on the surfaces of closed buildings, such as those described in *Explosive shocks in air*[24] and *Structural design for dynamic loads*[27].

It is important to be aware that structures within congested city centre environments will suffer from multiple reflections, diffraction, and confinement effects from neighbouring structures, which will add to the complexity of blast load prediction. In these cases, it is necessary to resort to the computational methods listed in Section 7.9.

7.8 Internal explosions in buildings

The detonation of a high explosive inside a building produces two types of loading:

- loading due to initial blast wave and due to subsequent blast waves caused by reflections
- loading due to explosion by-product expansion.

The confinement of an initial blast wave can cause a train of blast waves to be produced. The train of blast waves is caused by reflections, so blast overpressures stay high for longer.

At the same time, the gaseous products of the explosion that remain, due to the confinement, cause very high intensity pressures to be developed. The build-up of gas pressure is commonly termed a *quasi-static* explosion.

In many cases, internal explosions are not totally confined because non-structural elements will fail or the explosion will be vented through a series of rooms or via orifices in doors. In other cases, blast-relieving panels may be provided to relieve the overpressures.

Prediction of internal blast overpressures is difficult, however simplifications can be made to estimate the peak overpressure and the duration of the internal blast can be assumed (see Baker *et al.*[20]). These parameters enable a conservative estimate of the response of the enclosure to be determined.

If accurate predictions are required, it is necessary to resort to the use of computational techniques using software packages such as those listed in Section 7.9.

7.9 Computational methods for predicting blast overpressure and loads

Equations and charts relating the shock intensity, duration, and impulse have been developed since the 1800s and are still used today to make a quick estimate of the blast characteristics. Many of these equations are limited by their range of applicability, and the accuracy of all empirical equations diminishes as the explosion becomes increasingly *near field* (i.e. blast overpressures greater than 10 bar). As the prediction of accurate blast overpressures in real situations is a complex task, it is necessary to resort to computational techniques.

7.9.1 Methods available for prediction of overpressure

The following methods are available for the prediction of blast overpressures:

- empirical methods
- phenomenological methods
- numerical methods.

Empirical methods

These are essentially correlations with experimental data. Their ability to handle different scenarios stems from a simple interpolation between data points. Their limits of applicability and accuracy depend largely on the extent of the underlying experimental database. They are typically implemented as a computer program, and are simple and quick to use. They are, however, the most restricted methods, and certainly cannot address scenarios if there are major limitations in the underlying experimental data.

Phenomenological methods

These methods are more fundamentally based than are empirical methods. They attempt to model the underlying important physical processes, albeit in an approximate way. Their predictive accuracy will be better than that provided by the empirical methods. They provide the ability to interpolate more accurately between data and to extrapolate with more certainty to situations not addressed by experimental work.

Numerical methods

Numerical models are the most fundamentally based of all; they solve the underlying equations describing the behaviour of gases, combustibles, and solids. They are commonly termed computational fluid dynamic (CFD) models. Computational fluid dynamic models have the potential to provide a higher predictive accuracy and a greater potential of addressing any scenario. In practice, the accuracy is limited by the required computing power, the accuracy of the numerical methods, and the underlying physical sub-models. The models can be difficult to use and require a high level of user expertise.

Some of the programs used today are listed in Table 7.3.

Table 7.3 *A selection of software packages*

Program	Corporate author and notes
AUTODYN-2D and 3D	Century Dynamics Blast simulation, impact and penetration
FLACS	Christian Michelsen Research Blast simulation
BLASTIN	Applied Research Associates Inc.[28] Used for internal explosions
CHAMBER	Applied Research Associates Inc.[29] Effects inside buildings caused by external explosions
DYNA 3D	Lawrence Livermore National Laboratory Blast simulation in three dimensions[30]
BLASTX	Science Applications International Corporation[31] Blast in multi-form structures
CTH	Sandia National Laboratories[32]
FEFLO	Science Applications International Corporation[33]
FOIL	Applied Research Associates[34] Ground shock prediction
HULL	Orlando Technology Inc.[35]
SHARC	Applied Research Associates; Hikida *et al.*[36]
CONWEP	US Army Waterways Experimental Station[37][38]

Overpressure calculations should generally be conducted by a specialist who has access to experimental data to validate the method used.

8 DYNAMIC RESPONSE

This Section presents the types of response that can occur when a structure or structural component is impinged on by an explosion. Analysis methods are presented that can be used to determine the maximum deflection or a complete time-history of displacement. The analysis methods are as those presented in TM5-1300 *Structures to resist the effects of accidental explosions*[12], which are based on theory presented by Biggs[39]. The response of real distributed mass structures and structural elements is idealised using single-degree-of-freedom and multiple-degree-of-freedom methods, where either elastic or elasto-plastic techniques are considered. Computational techniques are also discussed.

8.1 Response regimes

Blast load impinging on a structure imparts energy to the structure, causing it to behave dynamically. How the structure (or a structural element) responds to blast loading is linked closely to the ratio between its natural period and the duration of the blast. Three response regimes can be produced, which are defined as:

- impulsive
- quasi-static
- dynamic.

The three response regimes are commonly quantified as shown in Table 8.1.

Table 8.1 *Identification of response types*

Loading regime	Based on natural period of the structure
Impulsive	$t_d/T < 0.4$
Dynamic	$0.4 < t_d/T < 2$
Quasi-static	$t_d/T > 2$

where t_d is the duration of the blast load
 T is the natural period of vibration of the structural element.

It is important to note that for a particular blast wave, response could be impulsive for one structure but quasi-static for another, because of the different natural period of vibration of each structure.

8.1.1 Impulsive response regime

In an impulsive response regime, duration of the blast load is very short compared with the natural period of the structural element. The duration of the load is such that the load has finished acting before the element has had time to respond to the extent that it would if an equivalent static force were acting. Due to inertial resistance of the structure, most of the deformation occurs after the blast load has passed.

When an impulse is delivered to a structure, it produces an instantaneous velocity change, where momentum is acquired and the structure gains kinetic energy.

Impulse is an important aspect of damage-causing ability of this type of blast, and may become a controlling factor in design in situations where the blast wave is of a relatively short duration.

8.1.2 Quasi-static regime

Where the duration of the blast load is much longer than the natural period of the structural element, the loading is termed quasi-static. In this case, the blast loading magnitude may be considered constant while the element reaches its maximum deformation. The response tends to that for an equivalent static force.

For quasi-static loading, the blast will cause the structure to deform while the loading is still being applied. The loading therefore does work on the structure, causing it to deform and acquire strain energy.

8.1.3 Dynamic response regime

In between the impulsive and quasi-static regimes, there is a more complicated, time-dependent regime, commonly called the dynamic regime. In this regime, the load duration is similar to the natural period of vibration of the structural element, and the duration of the load is similar to the time taken for the element to respond significantly. There is amplification of response above that which would result from static application of blast load.

For dynamic load response, the complete energy relationship is applied where the work done by the blast load is equal to the kinetic energy and strain energy imparted to the structure.

8.2 Analysis methods

Essentially three methods of analysis are available to calculate the response of a structure subjected to transient loads (see Figure 8.1). These methods are termed:

- single degree of freedom
- multiple degree of freedom
- approximate methods (energy methods and static analysis methods).

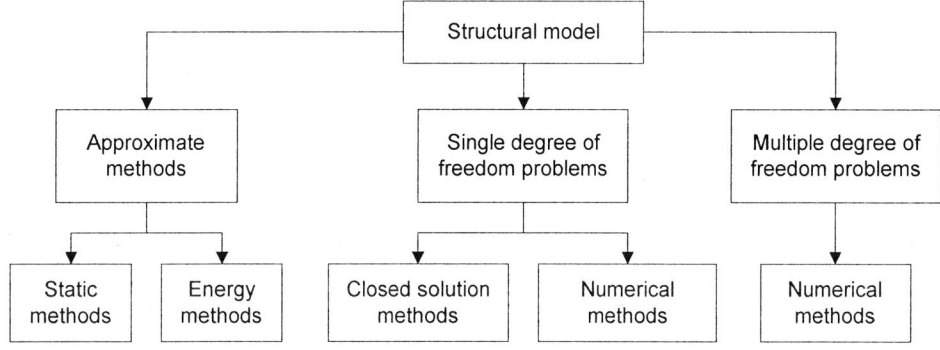

Figure 8.1 *Methods of analysis*

The merits of each of these methods are discussed briefly in Sections 8.2.1 to 8.2.3 and detailed considerations for each are given in Sections 8.3 to 8.5.

8.2.1 Single-degree-of-freedom method

Provided that a structural system can be idealised adequately to a single-degree-of-freedom system (a real system that is comparatively simple, e.g. a single plate or beam), this is the recommended approach for determining response to blast loading. This method is commonly carried out using closed solution or rigorous methods. Closed solutions, however, are limited to a small number of applications. Numerical analysis techniques may be employed to analyse single-degree-of-freedom problems, including elasto-plastic material behaviour as well as complex loading regimes.

The single-degree-of-freedom model has the ability to modify equations and parameters if a time-stepping procedure is employed. This enables a nonlinear system to be modelled, because the equations of motion can be modified to reflect plastic deformation and allow strain-rate effects to be included.

This method is most suited if the primary requirement in determining the behaviour of a blast-loaded structure is its final state (e.g. maximum displacement) rather than a detailed knowledge of its response history.

The principles of analysis for a single-degree-of-freedom system can be extended to specific structural elements that can be idealised as equivalent lumped mass structures by means of load and mass factors (see Section 8.3).

8.2.2 Multiple-degree-of-freedom method

Where a structure cannot be idealised to a single degree of freedom (e.g. more than one parameter is to be analysed), a more rigorous approach is required. This can be obtained by performing a multiple-degree-of-freedom analysis using numerical techniques (see Section 8.4).

Numerical analysis techniques are generally the most powerful and may be employed for systems of varying complexity. They may also be used for elasto-plastic material behaviour as well as complex loading regimes.

8.2.3 Approximate methods

Approximate methods are limited to energy methods and static analysis methods.

Energy methods

Energy methods are based on the principle that the work done by the applied load or the kinetic energy of the structure must be equal to the change in strain energy in the structure. They can be used for all load regimes. The energy methods are adequate for simple structural elements and load regimes, but for more complex structural elements or frames and load configurations these methods become very laborious and time consuming. They are not therefore recommended for any but the simplest cases (see Section 8.5).

Static analysis

Static-analysis methods have been used where quasi-static blast loads act (dynamic amplification in response is minimal). Both static-elastic and static-plastic analysis methods may be used where the magnitude of the load is assumed to be the peak value. As large conservatism in results can occur, these methods are not generally recommended.

8.3 Single-degree-of-freedom method

The dynamic behaviour of a structure (or structural element) can be reproduced reasonably accurately by idealising the structure as an equivalent single-degree-of-freedom model. The single-degree-of-freedom model reduces the structure to one comprising a concentrated mass joined by a weightless damped or undamped spring, which is subjected to a time-varying concentrated load (Figure 8.2).

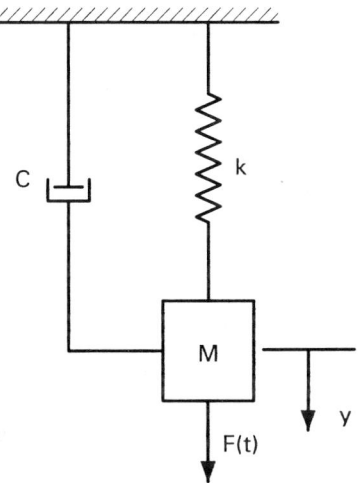

Figure 8.2 *Single-degree-of-freedom system*

8.3.1 Fundamental equation of motion

The fundamental equation of motion for the spring system that idealises the structure or structural component is given by:

$$M\ddot{y} + c\dot{y} + ky = F(t) = F\left(1 - \frac{t_d}{T}\right)$$

where $\ddot{y}$, $\dot{y}$, and y are acceleration, velocity, and displacement respectively
M   is the mass
c   is a damping coefficient
k   is the stiffness of the spring
F(t) is the load-time variation as defined in Section 7.6.

For blast-response analysis, the damping component is omitted because in most situations damping has very little effect on the fundamental response peak. Also, where plasticity is assumed, the energy dissipated through plastic deformation will be significantly greater than that dissipated by structural damping. By neglecting the damping term, a conservative analysis is adopted.

8.3.2 Equivalent lumped-mass systems

The single-degree-of-freedom analysis method can be applied readily to determine the response of more complex structures and load configurations by representing the structure as a single-degree-of-freedom equivalent lumped-mass system. Although a structure possesses many degrees of freedom, one degree of freedom is adequate because the fundamental mode of response usually predominates in the response of the structure to short duration blast loads. Table 8.2 gives examples of systems and loadings that can be represented by the spring system in Figure 8.2 and analysed by the single-degree-of-freedom method.

Table 8.2 *Systems suitable for single-degree-of-freedom analyses*

System type	Fixity	Loadings
(beam)	Pinned Fixed Propped cantilever Cantilever	Point Distributed
(plate)	Pinned Fixed	Pressure
(frame)	Pinned Fixed	Point Sway

Transformation factors

Actual structures are converted into idealised or equivalent dynamic systems by applying transformation factors to the dynamic parameters of the structure, i.e. to the mass, load, and stiffness of the real system. In this way, the displacements of the single-degree-of-freedom model are equivalent to the real system. The factors are theoretically derived, based on an assumed deflected shape of the element.

Transformation factors that are used to define an equivalent single-degree-of-freedom model of the structure include:

- mass factor K_m
- load K_L
- stiffness K_S
- load-mass factor K_{LM}

where

$$K_M = \frac{M_E}{M} \quad ; \quad K_L = \frac{F_E}{F} \quad ; \quad K_S = \frac{K_E}{K} \quad ; \quad K_{LM} = \frac{K_M}{K_L}$$

where M_E, F_E, and K_E are the equivalent mass, blast load, and stiffness of the system respectively, and K_S is approximately equal to K_L.

Transformation factors are presented in Tables C.1 to C.3 in Appendix C for a number of different beam and loading conditions taken from charts by Biggs[39]. There are a number of different strain ranges in these tables, which account for the change in beam response from elastic through elasto-plastic to plastic.

Natural period

The effective natural period of vibration T is required to determine the maximum response of the structure y_m and is given by:

$$T = \frac{2\pi}{\omega} = 2\pi\sqrt{\left(\frac{M_E}{K_E}\right)} = 2\pi\sqrt{\left(\frac{K_{LM}M}{K}\right)}$$

where ω is the natural circular frequency
M is the actual mass of the structure
K is the stiffness of the structure.

Dynamic reactions

Blast loads are variable with time and can produce reaction forces that are greater than the reaction forces that would be obtained if the loads were static.

Relationships to determine dynamic reactions V for various beams/slabs for elastic, elasto-plastic, and plastic states are presented in Tables C.1 to C.3 in Appendix C, where

F is the applied load
S is the static load needed to cause the same deflection as the blast load F.

8.3.3 Equation of motion for the equivalent single-degree-of-freedom model

The equation of motion for the equivalent lumped-mass single-degree-of-freedom model can be rewritten to include the transformation factor, i.e.

$$K_{LM}M\ddot{y} + R(y) = F\left(1 - \frac{t_d}{T}\right)$$

where the spring resistance term ky is replaced by a general resistance function $R(y)$.

Resistance function

The resistance of a structure (beam or slab) is a measure of the beam's resistance during particular response regimes. Depending on the structure constraints, there will be a resistance term associated with elastic, elasto-plastic, and plastic conditions. The maximum resistance that a structure can develop is given by the ultimate resistance R_m. For the single-degree-of-freedom analysis method, a simplistic bilinear resistance-deflection function is assumed, as shown in Figure 8.3. The displacement at the elastic limit (where R_m is reached) is y_{el} and the maximum displacement attained is y_m.

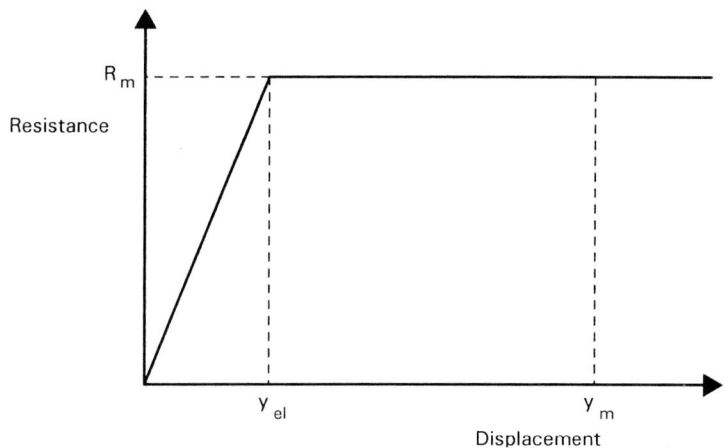

Figure 8.3 *Idealised resistance-deflection curve*

For structural components with rigid connections, elastic and elasto-plastic responses resistance will occur prior to ultimate resistance being attained. These more complex resistance functions can be simplified to an equivalent resistance function as shown in Figure 8.4. Both the original and the simplified resistance functions maintain equivalency because strain energy is preserved. The suffixes e, ep, and p relate to the elastic, elasto-plastic, and plastic regimes, while suffix E relates to the equivalent resistance function.

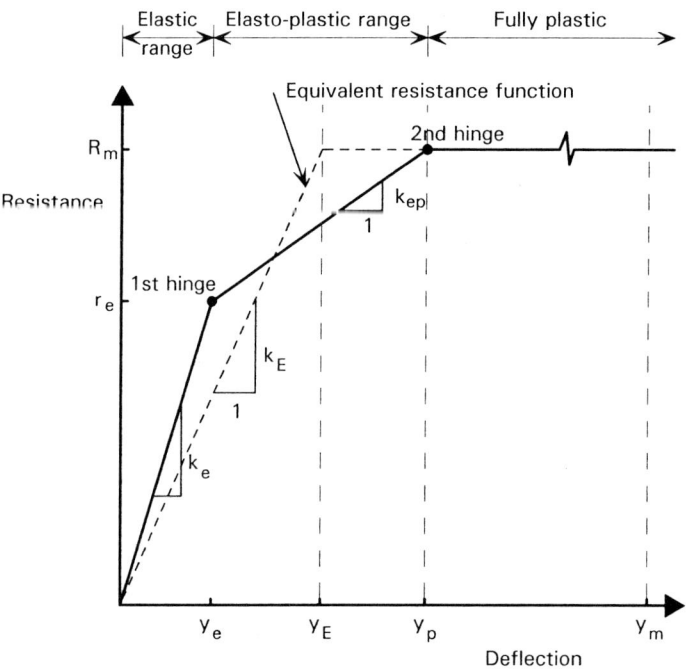

Figure 8.4 *Trilinear resistance function and equivalent bilinear resistance function*

The maximum resistance for various beams/slabs for elastic, elasto-plastic, and plastic conditions is also given in Tables C.1 to C.3 in Appendix C.

It is assumed that rotation of the first hinge can take place without local or lateral buckling taking place.

8.3.4 Response

In most cases, only the maximum displacement of a component rather than its response as a function of time is required. The maximum response of a single-degree-of-freedom model subjected to an idealised blast load is presented in the form of non-dimensional curves (see Figures C.1 to C.3 in Appendix C). To utilise these response charts, both blast load (pressure-time history) and the resistance-deflection curve of the structure are idealised to linear functions.

The response of a structure subjected to a dynamic load is defined in terms of its maximum deflection y_m and the rise time t_m to reach this maximum deflection. The dynamic load is defined by its peak value F and duration t_d, while the single-degree-of-freedom model is defined in terms of its ultimate resistance R_m, elastic deflection y_{el}, and natural period of vibration T. Response charts relate the dynamic properties of the blast load (F and t_d) to those of the element y_m/y_{el} and t_m/T; these are plotted as a function of R_m/F and t_d/T.

Elasto-plastic regime

Response charts for a zero rise time triangular idealised load-time profile in the elasto-plastic regime are presented in Figures C.1 and C.2 in Appendix C. Response charts for other load-time profiles can be obtained from *Introduction to structural dynamics*[39] or any other good book on structural dynamics.

It should be noted that the curve $R_m/F = 2$ represents completely elastic response. If $R_m/F > 2$, reference should be made to the elastic response regime below.

Elastic response regime

To obtain the response in the elastic regime, it is necessary to define the concept of the dynamic load factor (DLF) given by:

$$DLF = \frac{y_m}{y_{static}}$$

where y_m is the maximum dynamic deflection
  y_{static} is the deflection that would have resulted from the static application of the peak load F.

Because deflections, spring forces, and stresses in an elastic system are all proportional, the dynamic load factor may be applied to deflections, forces, and stresses to determine the ratio of dynamic to static effects.

The elastic response charts are presented in Figure C.3 in Appendix C.

8.4 Multiple-degree-of-freedom method

Multiple-degree-of-freedom systems such as single- and multi-storey frames are based on lumped-mass assumptions where the number of degrees of freedom is equal to the number of types of motion possible within a system, and not necessarily by the number of lumped masses. For each degree of freedom, there is a corresponding equation of motion. These equations of motion are used to determine the natural frequencies of a structure and corresponding characteristic shapes. Although several methods of analysis are available, in practice finite-element analysis techniques provide the most acceptable level of accuracy for the dynamic response of structures.

Finite-element analysis is a technique used to determine the dynamic response of a structure under the action of any general time-dependent loads. This type of analysis can be used to determine the time-varying displacements, strains, stresses, and forces in a structure as it responds to any combination of loads. Hence, finite-element methods require the use of more sophisticated software and specialist skills. The type of problems that warrants the use of finite-element methods usually requires intensive modelling input, and computer run times can be lengthy.

The advantage of the finite element method is its flexibility, and, unlike the single-degree-of-freedom method, it is unnecessary to assume a deformed shape or to idealise support conditions. Additionally, there may be analysis features that are unavailable with other methods, such as strain-rate effects, strain hardening, and temperature variable properties.

Care should be taken in analysing results obtained from finite-element methods, particularly to confirm the accuracy of the solution. Comparisons can be undertaken by an independent confirmatory finite-element analysis, preferably using a different program, or approximately by use of energy or single-degree-of-freedom methods.

Structural design for dynamic loads[27] gives further information.

8.4.1 Commercial computer analysis programs

There are two fundamentally different computer analysis programs: *uncoupled* and *coupled*. Each is discussed briefly.

Uncoupled computer analysis programs

Uncoupled computer analysis programs are programs in which there is no interaction between the blast loading and the response of the structure, e.g. ABACUS, ANSYS, AUTODYN, DIANA, DYNA-3D, EPSA-II, FLEX, LUSAS, and NASTRAN.

Coupled computer programs

A complication of dynamic response is that the very motion of the disturbed structure creates an alteration in the applied pressure. If this effect has to be accounted for, a more complex calculation has to be undertaken in which the two effects are coupled together. In coupled calculations, the computational fluid dynamic model for blast-load prediction is solved simultaneously with the computational solid mechanics model for structural response. By accounting for the motion of the structure while the blast calculation proceeds, the pressures that arise due to motion and failure of the structure can be predicted more accurately. Some coupled computer analysis programs are identified in Table 8.3.

Table 8.3 *Coupled computer analysis programs*

Name	Corporate author	Reference
ALEGRA	Sandia National Laboratories	Budge and Perry[40]
ALE3D	Lawrence Livermore National Laboratory	American Society of Mechanical Engineers[41]
DYNA3D/ FEFLO	Lawrence Livermore National Laboratory/SAIC	Lohner *et al.*, 1995[42]
MAZE	TRT Corporation	Schlamp *et al.*, 1995[43]

8.5 Energy methods

Energy methods can be used readily to analyse response due to blast loading for simple structural members (beams) in isolation. Analysis can be undertaken for both elastic and elasto-plastic regimes where the blast loading may be impulsive or dynamic/quasi-static. Response using the energy methods, however, is restricted to predicting maximum values, i.e. maximum displacement, maximum stresses, and maximum strains.

Referring to Section 8.1. and Table 8.1, the impulsive and quasi-static energy relationships can be represented mathematically for the elasto-plastic regime by:

$$\frac{I^2}{2M_E} = \frac{1}{2} y_{el} R_m + R_m(y_m - y_{el})$$

and for the impulsive regime by

$$F y_m = \frac{1}{2} y_{el} R_m + R_m(y_m - y_{el})$$

where I is the impulse
  M_E is the equivalent mass (see Section 8.3.2)
  F is the force
  R is the resistance
  y is the displacement

and the blast load is assumed to be triangular with zero rise time and the resistance of the member is as shown in Figure 8.3.

If required, standard text books such as *Introduction to structural dynamics*[39] are available, which give solutions for typical structural elements. Although the publication by Biggs is now out of print, other publications provide an excellent treatise of the subject; one such publication is *Theory of vibrations with applications*[44].

It is important to note that for a particular blast load, response may be impulsive for one type of structure while quasi-static for another. This would occur where the natural period of vibration of each structure differed significantly.

8.6 Pressure-impulse diagrams

It is not always necessary to assess structural response mathematically from first principles. Instead, the behaviour of any system that has been subjected to blast loading (e.g. a structure or a human being) can be represented diagrammatically by pressure-impulse diagrams. These diagrams are useful aids, which can readily predict levels of damage for the complete spectrum of load-impulse combinations.

Pressure-impulse diagrams are obtained readily by considering and representing the behaviour of a system over the whole range of possible response regimes. This is achieved by examining the behaviour of the system at the extremities of response, one extremity being the impulsive response regime and the other the quasi-static response regime. In each instance, the energy of the system is considered.

In the impulsive-response regime, it is assumed that the blast load is of such a short duration that the load has finished acting before the system has had time to respond significantly. Therefore in the limit, this type of load produces an instantaneous velocity change to the system. The behaviour is thereby determined by equating kinetic energy and strain energy of the system.

At the other extremity, in a quasi-static response regime, the blast load is of such a long duration that it remains constant while the structure attains its maximum deflection constrained by the strength of the structure. In the limit, the work done by the force is equated to the strain energy of the system.

The pressure and impulse relationships produced enable iso-damage curves to be plotted graphically based on predefined damage criteria. The pressure and impulse relationships can also be rearranged in non-dimensional form to produce specific impulse and pressure.

These pressure-impulse diagrams can be based on analytical, experimental, or real-life events. To show how these pressure-impulse curves can be used, various real-life cases appropriate to damage to structures and people are presented. A summary of the cases presented in *Blast and ballistic loading of structures*[5] is included.

8.6.1 Safe stand-off distances for explosive testing

An iso-damage curve to determine safe stand-off distances for explosive testing is shown in Figure 8.5.

This curve is based on a study of houses damaged by bombs dropped in the UK during the Second World War[45]. The categories A, B, C_b, C_a, and D define levels of damage that can be used not only for houses but also for small office buildings and light-framed factories. They are defined as:

Category A Almost complete demolition.

Category B Such severe damage as to necessitate demolition: 50-60% of external brickwork destroyed or unsafe.

Category C_b Damage rendering house temporarily uninhabitable - partial collapse of roof and one or two external walls. Load-bearing partitions severely damaged and requiring replacement.

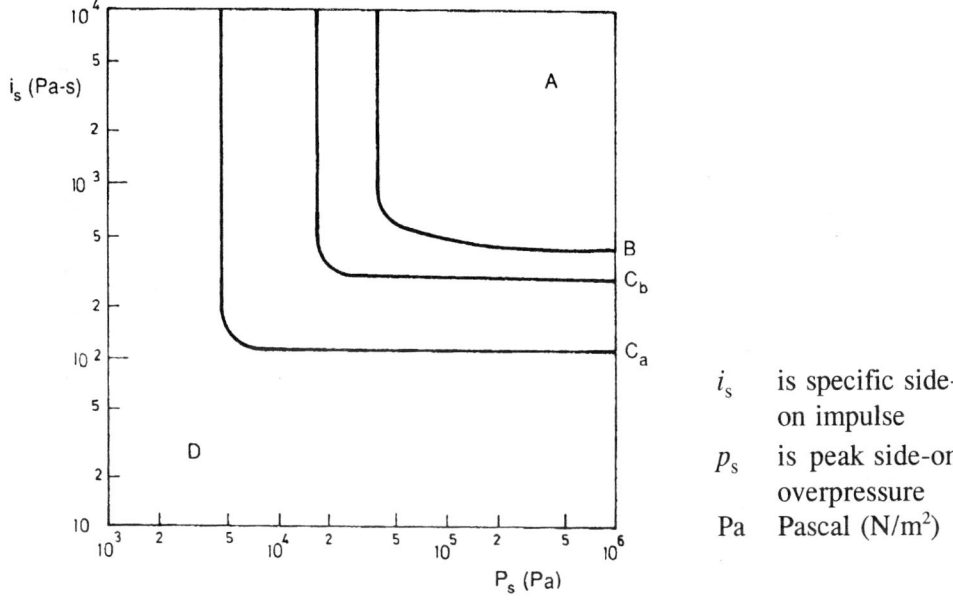

Figure 8.5 *Pressure-impulse diagram for damage to houses*[20]

Category C_a Relatively minor structural damage yet sufficient to make house temporarily uninhabitable. Partitions and joinery wrenched from fixings.

Category D   Damage calling for urgent repair but not so as to make house uninhabitable. Damage to ceilings and tiling. More than 10% of glazing broken.

By superimposing range-charge mass overlays on the pressure-impulse diagram above, the damage to buildings caused by specific explosive devices can be assessed readily. This relationship between blast and scaled distance is shown in Figure 8.6[16][36]. GP250 and GP2000 are general purpose bombs of mass 118 kg and 895 kg respectively (TNT equivalent mass 55 kg and 542 kg respectively), which are based on detonation occurring at the ground surface. In addition, 2.2 tonne TNT and 10 tonne TNT devices are presented, which have been assumed to be detonated on the ground.

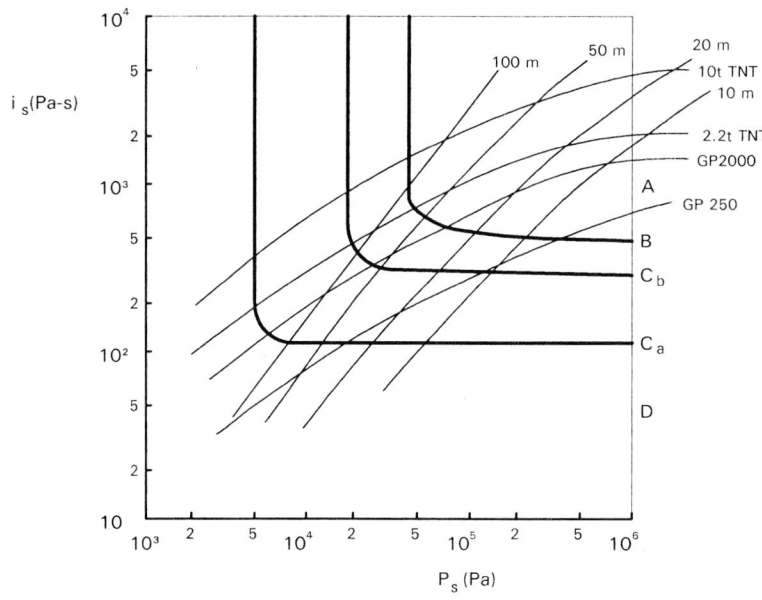

Figure 8.6 *Pressure-impulse diagram for damage to houses with range-charge mass overlay*

Inspection of the graph above shows that for 10 tonne of TNT at 100 m range, almost complete demolition of a dwelling occurs, while only damage rendering a house temporarily uninhabitable occurs for 2.2 tonne TNT at the same range.

8.6.2 Human response to blast loading

Pressure-impulse diagrams also allow prediction of personnel injury from blast loading. Lung and ear damage are considered here.

Lung damage is caused directly by the blast wave impulse being induced in the body. Figure 8.7 is a pressure-impulse diagram showing threshold level of damage and survivability indices for more severe damage.

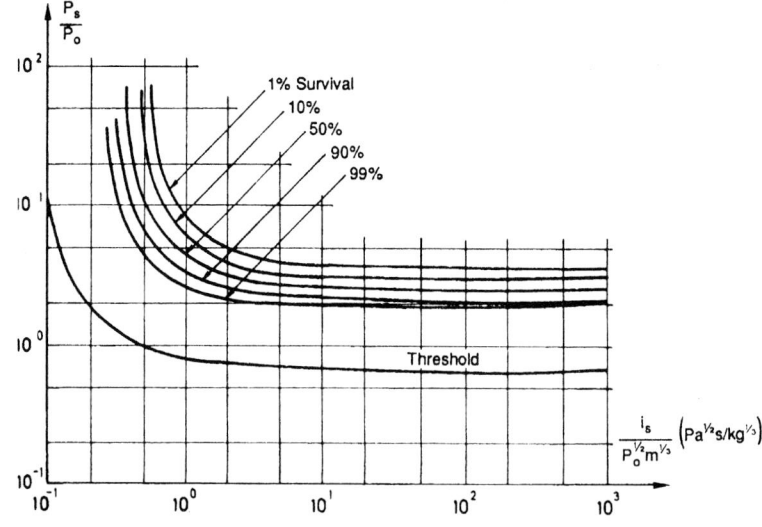

Figure 8.7 *Pressure-impulse diagram for lung damage in humans*[20]

Figure 8.8 presents a pressure-impulse diagram showing threshold levels for ear damage caused directly by the blast wave. Parameter *m* is the mass of the subject and label TTS_1 represents the temporary threshold shift that is characterised by temporary deafness.

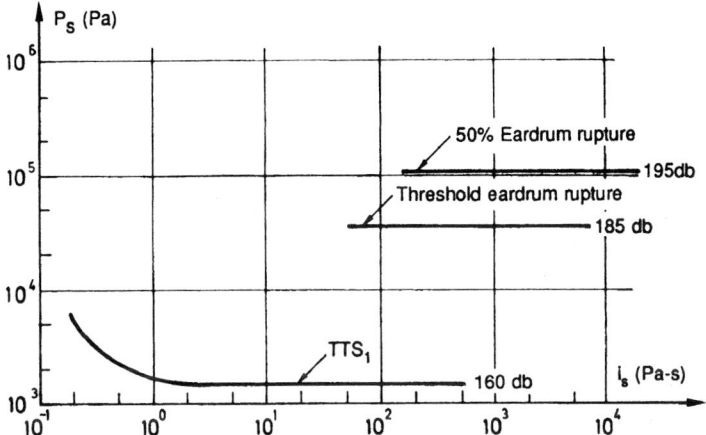

Figure 8.8 *Pressure-impulse diagram for ear drum rupture*[20]

For further information specific to human injury caused by blast load, reference can be made to research carried out by Bowen *et al.*[46] and Ahlers[47]. Summaries of these results can be obtained from Baker *et al.*[20][48]

8.6.3 Glazing behaviour

The complex behaviour of glazing in resisting blast loading can also be represented simply but comprehensively using pressure-impulse curves. As an example, a pressure-impulse diagram is presented showing the behaviour of a laminated glass specimen against blast[49]. The specimen is 15.4 mm thick and plan size 1.5 m² (Figure 8.9).

The left-hand curve corresponds to cracking a sheet of glass acting as a plate in bending, whereas the right-hand curve represents the equivalent membrane action of the lamination. The dotted diagonal lines indicate the duration of the blast. As the blast duration increases (from 10 to 40 ms), the blast overpressure that the specimen can resist reduces significantly from 1 to 0.4 bar.

Appendix D gives further information on laminated glass.

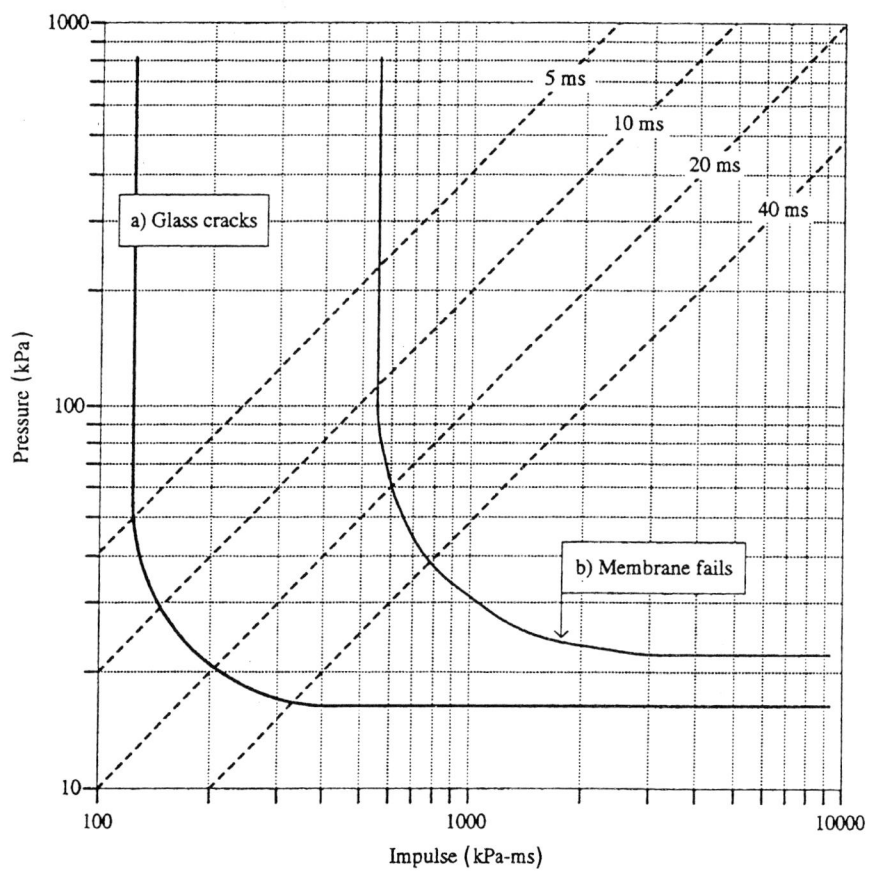

Figure 8.9 *Schematic pressure-impulse curves for a laminated glass pane*

9 MATERIAL PROPERTIES

The ability of a material to resist blast depends on the strength and the ductility of the material. This Section provides additional material properties, which are not required for static analysis but are fundamental when assessing the behaviour of a structural component when it is acted upon by blast loading. Additional material property information is required because:

- the normal static stress-strain relationship is altered by strain-rate effects, permitting different levels of deformation and energy absorption to take place
- the dynamic loading may affect the circumstances under which brittle failure can occur.

The materials commonly used in building components are presented. These include steel, concrete, and stainless steel.

9.1 Behaviour of materials under dynamic loading

The mechanical properties of materials are affected by the rate at which straining takes place. Those materials having definite yield points and pronounced yielding zones show a marked variation in mechanical properties with changes in loading rate. Yield strengths are generally higher under rapid strain rates than under slowly applied loads. A large part of the increase in strength with rapid strain rate is attributed to a lesser amount of plastic deformation having time to occur during dynamic loading. A higher stress is therefore required to produce a failing strain.

In an actual structure, the strain rate is determined by the response of the structure to dynamic loads. A dynamic analysis can provide estimates of maximum strain rates, which will generally vary with time and location in the structure. Because structural response, strain rate, and yield strength are interdependent, a trial-and-error method of analysis is necessary, however in view of uncertainties in other variables, great precision in evaluation of strain-rate effects is not normally justified. Based on this uncertainty, the strain-rate effects for carbon steels are estimated based on the time of the structural component to reach yield stress. This is based on the work performed by Newmark and Rosenblueth[50]. A more general measure of strain rate is quoted for stainless steels and concrete.

9.2 Strain-rate effects in structural steels

The mechanical properties of the standard low carbon structural steels (S275 and S355 as specified in BS EN 10025[51] and BS EN 10210[52]) are affected noticeably by the rate at which straining takes place. If the mechanical properties under static loading are considered as a basis, the effects of increasing strain rate can be illustrated in Figure 9.1 and these can be summarised as follows:

- the yield point increases substantially to a dynamic yield stress value
- the modulus of elasticity generally remains insensitive to the rate of loading
- the ultimate tensile strength increases slightly, however the percentage increase is less than that for the yield stress

- the elongation at rupture either remains unchanged or is slightly reduced due to the increased strain rate.

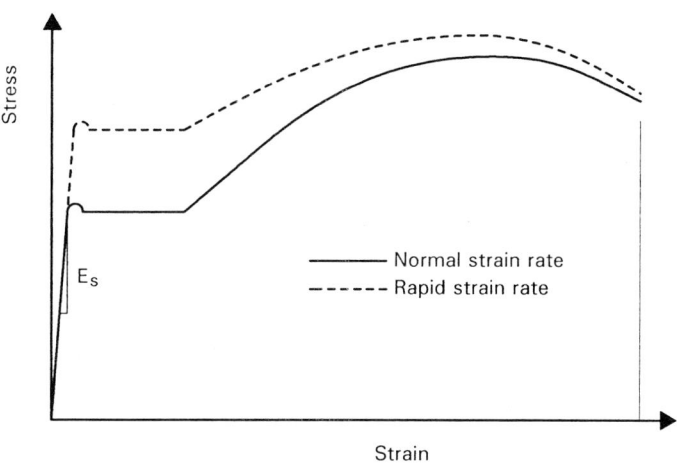

Figure 9.1 *Typical stress-strain curve for a common structural steel*

9.2.1 Dynamic increase factor

The enhancement in stress (yield or ultimate) due to dynamic strain-rate effects is determined from knowledge of the dynamic increase factor (DIF) where:

$$DIF = \frac{\sigma_{dyn}}{\sigma_y}$$

and σ_{dyn} is the dynamic yield stress corresponding to a particular strain rate
σ_y is the yield stress under static conditions (no strain-rate effects).

Dynamic increase factors to be used for grades S275, S355, S460, and reinforcement steels are shown in Table 9.1. The dynamic increase factors are based on the time to reach yield stress in the steel member.

Table 9.1 *Dynamic increase factors for structural steels*

Time to reach yield stress	Dynamic increase factor for yield stress[1]	Dynamic increase factor for ultimate yield strength[2]
>1 s	1	1
100 ms	1.1	1.05
10 ms	1.6	1.05
1 ms	1.9	1.05

Note:
1. For the higher strength steels, i.e. S460 and reinforcement steels[53], yield exceeds 430 N/mm². As these higher yield steels do not have a definite yield point or pronounced yielding ranges, *proof stress* values at 0.2% or 0.5% proof strain are commonly used.
2. These dynamic increase factors only apply to S275 and S355 steels. For the higher yield steels, the dynamic increase factor is equal to 1.0 throughout the strain range.

9.2.2 Dynamic design yield stress

The dynamic design yield stress $\sigma_{y,des}$ for bending is given by

$$\sigma_{y,des} = a \, (DIF) \, \sigma_y$$

where a is a factor that takes into account the fact that the yield stress of a structural component is generally higher than the minimum specified value given in BS 5950. For S275 and S355 steels, a = 1.10.

9.3 Strain-rate effects in stainless steels

Austenitic stainless steels are highly ductile materials (twice as ductile as mild steel) that have superior energy-absorbing capabilities. They have a strong strain dependency, where as strain rate is increased, their strength (at 0.2% strain) increases noticeably, while their rupture strain is decreased. Hence they are an attractive material for blast walls and other structural items where blast resistance is to be provided.

Austenitic stainless steels to EN 10088-2[54] include Grade 1.4307 (commonly known as 304L) and Grade 1.4404 (316L). Typical stress-strain plots for a 304L stainless steel at room temperature for a range of strain rates are presented in Figure 9.2.

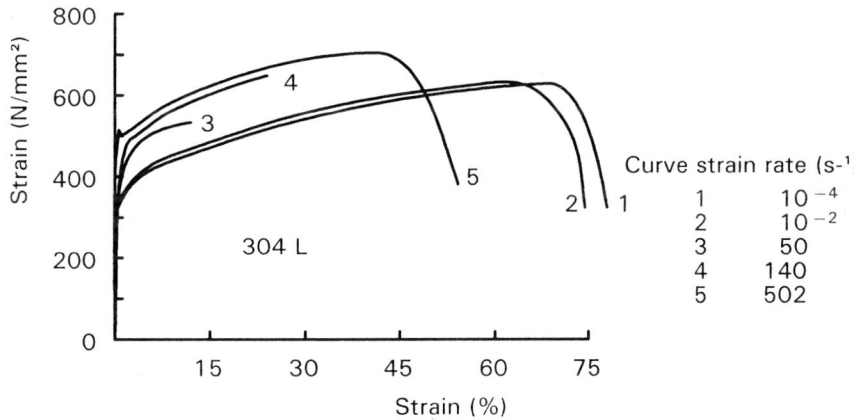

Figure 9.2 *Strain rate effects on a 304L stainless steel*

Strain-rate effects for 316L austenitic stainless steel are presented in *Design of stainless steel blast walls*[55] and shown in Table 9.2. No data are presented for 304L stainless steel, however strain-rate effects for these two stainless steels are similar.

The strain-rate enhancement for ultimate strength of Grade 1.4307 stainless steel is not as pronounced as at 0.2% proof strain. At a strain rate of 212 s^{-1}, the dynamic increase factor is only 1.07.

Table 9.2 *Dynamic increase factors for Grade 1.4307 (316L) stainless steel at 0.2% proof strain*

Strain rate (s^{-1})	Dynamic increase factor (0.2% proof strain)
2.50×10^{-3}	1
8.63×10^{-3}	1.04
1.78×10^{-2}	1.07
8.80×10^{-2}	1.13
7.42	1.28

9.4 Strain-rate effects in concrete

Concrete is a brittle material in comparison with structural steel and reaches its maximum strength at strains near the yield point of some reinforcing steels. Also its stress-strain relationship is nonlinear over most of its working range.

Among the many variables affecting the compressive strength of concrete are the types of material constituents, their proportions, the method and time of curing, and age. Most of these variables are accounted for by specifying a 28-day strength of test specimens, however the strength of a concrete mix normally continues to increase beyond its 28-day strength. While it is conservative to neglect this increase for design purposes, the actual strength of concrete may be ≥ 50% greater than that specified at 28 days.

The unconfined compressive strength of concrete increases with the rate of loading, and for very high rates of strain, the dynamic ultimate strength increase factor can be very much greater than the static ultimate strength (i.e. 20-40% increase).

For reinforced concrete elements used in composite design, Table 9.3 provides dynamic increase factors for concrete.

Table 9.3 *Dynamic increase factors for concrete*

Mode	Dynamic increase factor
Bending	1.25
Shear	1.1
Compression	1.15

10 STRUCTURAL DESIGN APPROACH

In design where checks are to be performed to assess the ability of the building to withstand explosions, the elastic limit of structural components is often exceeded by a considerable margin and elasto-plastic behaviour has to be considered. It is therefore necessary to impose additional requirements to those stated in the Codes of Practice presented in Section 5.2. The main aim of these requirements is to guarantee energy-absorption capability by appropriate detailing. Above all, this is to facilitate controlled distortion under overload without premature loss of strength by instability. It has to be emphasised that acceptance of finite distortion in a structure subject to blast loading is an admissible design approach, and an informed choice about the level of damage deemed acceptable has to be made at the beginning of the design process.

A procedure is also presented in Section 10.3 showing how the response of a structural component can be determined and whether it satisfies the damage acceptance criteria.

10.1 Design requirements

The following design requirements are to be satisfied:

- A structure is to be capable of withstanding an explosion of realistic magnitude in order to protect personnel, and equipment it houses, from the damaging effects of the blast.

- In the event of an explosion of damage potential in excess of the design values, the structure shall deform without any significant loss of its load-carrying capacity, thereby providing an adequate margin of safety against catastrophic collapse.

- The building shall safely support the post-explosion conventional design loads with some minor repairs.

10.1.1 Design approach

The *design approach* is the combination of an analysis method with a means of strength assessment, which is usually a design code. Strength and deformation criteria relevant to explosion events may be set on an individual basis, but will in general reflect the design philosophy adopted.

For design against explosion loading, the following factors are to be considered:

- the time-varying nature of the loading
- the enhanced yield strength of steel at high strain rates
- response can take place outside the linear elastic region commonly used for in-service design conditions
- large displacements of structural components may be permitted, providing that progressive collapse is not encountered
- damage-acceptance criteria
- reduction in the intensity of other loads.

The ultimate limit state approach to BS 5950-1 is used to demonstrate that a structure can resist an explosion, with or without damage. The underlying philosophy permits plasticity at the ultimate limit state. Indeed, controlled rotation of plastic hinges is an excellent way of absorbing blast energy. In order to justify controlled (but finite) rotation through significant angles, cross-sections need to be proportioned appropriately and restrained laterally. Connections must be strong enough to develop the plastic capacities of members expected to yield. The degree to which plastic hinges can rotate generally has to be assessed experimentally.

10.1.2 Robustness and disproportionate collapse

The clauses in BS 5950: *Structural use of steelwork in buildings* that relate specifically to robustness and disproportionate collapse as stated in the Building Regulations are discussed in BS 5950-1: Clause 2.4.5[56]. Conservatively, these code measures impose an inherent level of blast resistance on all modern structures.

In BS 5950-1, it is a requirement that all steel structures, irrespective of height or span, comply with prescribed minimum acceptance levels of robustness. Clause 2.4.2.3 *Sway stability* prescribes that all structures should be able to resist notional disturbing forces dictated by the design gravity loading. Clauses 2.4.5.1 and 2.4.5.2 specify *robustness-specific* requirements that must also be satisfied by all structures.

Essentially, Clause 2.4.5 ensures that the members of even the most modest structures are tied together adequately in all directions. Structures that fall outside the restriction on the number of storeys or span (in the case of public buildings) must be appraised with due consideration of their behaviour under collapse conditions.

There are effectively three alternative design routes available - namely, the *tying force route*, the *localisation of damage route*, and the *key element route*.

Tying force route

The tying force route is the preferred route where the philosophy of providing a positive tie between intersecting members is fundamental to ensuring adequate resistance to disproportionate collapse. In addition, it is considered that compliance with this approach requires the minimum amount of design effort. Using this approach, a robust design can be achieved by satisfying a number of requirements regarding the location of principle bracing elements for sway resistance, tensile resistance of column splices, anchoring of floor units, and the provision of peripheral and internal ties. There is a proviso, however, that beams that support columns (e.g. transfer structures) should be treated as *key* elements.

The first of these conditions limits the reliance of large parts of the structure on any single system of sway bracing. Obviously, in circumstances where such a bracing system is rendered ineffective by a localised blast or accidental loading, the consequences on a pure pin-jointed structure could be catastrophic. It should be noted, however, that most nominally-pinned structures (regarded as simple joints) do possess some inherent moment capacity in the connections, and this may be taken into account when assessing the resistance to lateral loads under accidental conditions.

A BCSA/SCI publication *Joints in simple construction*[57] provides a comprehensive summary of the failure loads that can be achieved for a range of connection types, and provides a design method for predicting the tensile resistance of connections based on large deformation theory.

Localisation of damage route

An alternative to providing enhanced resistance to tying forces is to appraise the building in accordance with the requirements for localisation of damage. Structures should be inherently capable of limiting the spread of local failure regardless of the cause and ideally should be capable of locally bridging over a missing member - albeit in a substantially deformed condition. In this case, the missing member can be any single column or beam carrying a column. If it is found that removal of a member results in damage more extensive than the specified limits, the member must be designed as a *key element*.

Key element route

Adopting this approach, any member on which significant proportions of a structure rely for stability and support can be designed as a *key element*. The key element is designed to resist abnormal loadings that might otherwise render it ineffective.

If the absolute safety and integrity of a structure is to be guaranteed, key elements will have to be designed to resist all foreseeable, abnormal loading conditions. Clearly, to consider the consequences of all these conditions would be highly subjective and result in an uneconomic design. At present, there is no guidance in BS 5950-1 on the magnitude of abnormal loads that should be applied, however in the proposed revision of BS 5950-1, i.e. Draft Document 98/102164[58], it is recommended that accidental loading as specified in BS 6399-1[26] should be used. The abnormal loading that has been used in the past has been the *infamous* 34 kN/m^2 pressure (blast loading) as stated in the Building Regulations.

Further detailed information on the implementation of robustness to steel structures and recommended modifications to requirements for structural integrity are included in SCI Report SCI-RT-204 1992 (confidential).

10.1.3 Current European proposals

ENV 1991-2-7[10] on accidental actions due to impact and explosions has now been published as a European Prestandard, and a National Application Document is being drafted for trial use in the UK. The ENV proposes three approaches to design for accidental actions, each assigned to a different category of accidental design situations.

Category 1   defined as having *limited consequences*, requires no specific considerations for accidental loads.

Category 2   with *medium consequences*, requires either a simplified analysis by static equivalent models or the application of prescriptive design/detailing rules, depending on the specific circumstance of the structure in question.

Category 3   relates to *large consequences*, recommending a more extensive study, using dynamic analyses, nonlinear models, and load-structure interaction if considered appropriate.

The application of these requirements will be a matter for the relevant national regulatory authorities.

10.1.4 Load and resistance factors for design

Load factors

The design events identified by the techniques described in Section 3.2 are extreme events. If they have been identified by engineering judgement as being worst case credible scenarios, they are likely to have a very low probability of occurrence.

In determining the best estimate of blast overpressure loads, it is necessary to recognise the imprecision of overpressure predictive techniques.

The following load factors are suggested for general adoption. Other factors may be adopted at the discretion of the engineer in particular circumstances.

- Dead loads and other permanent loads should be factored by 1.0.

- Variable imposed loads should be set at 1.0 times best estimates of likely load immediately prior to an explosion. This load is likely to be only a small proportion of design load. In similar circumstances, BS 5950-1 proposes a factor of 0.33 times design load and this figure is put forward in the absence of more specific guidance.

- Blast loads should be factored by 1.0.

- Environmental loads can be ignored when considering significant blast loading.

- Internal loads controlled by plasticity: a particular problem arises where the reactions from a substructure onto its supporting structure are controlled by plasticity in the former. Account needs to be taken of the likelihood that the material going plastic has a greater than minimum strength. In the absence of more specific information, it is suggested that these internal forces are factored by 1.2 when considering their effects on the supported structure. Alternatively, and more appropriately for simple components, bounding analyses can be carried out with upper and lower bound estimates of the yield stress and the design based on the resulting envelope of forces.

Resistance factors

For response of structural steelwork, it is appropriate to recognise enhancement of the strength from strain-rate effects where the elements are sufficiently robust and compact for their capacity not to be reduced by local or overall buckling. It is also necessary to take into account that the yield stress of a structural component is generally higher than the minimum specified value given in BS 5950, especially for connection design.

The following resistance factors for structural steelwork are suggested for general usage. Other factors may be adopted at the discretion of the engineer in particular circumstances.

- The measured yield stress can be used if it is known.

- Account may be taken of strain-rate effects in accordance with Section 9.

- An enhancement factor of 1.1 is to be applied to yield stress to take into account that the yield stress of a structural component is generally higher than the minimum specified value given in BS 5950 (see Section 9.2.1).

10.2 Acceptance criteria

It is necessary to define a criterion that is to be used to assess the performance of a structure subjected to blast loading. As an explosion is an extreme event, the acceptance criteria adopted may differ from those used for conventional design.

The main acceptance criteria currently used in design are:
- strength limit
- deformation limit.

10.2.1 Strength limit

Where strength governs design, failure is defined as occurring when the design load or load effects exceed the design strength in a manner that is similar to conventional design.

The criterion may be applied in the elastic as well as plastic regimes. The only difference for the explosion case is that modified factors on loading and/or strength may be adopted in recognition that it is an extreme event.

10.2.2 Deformation limit

In some cases, acceptance of a degree of permanent deformation may not only be tolerated but may be an essential feature of the design. This criterion may therefore be applied in a markedly different way for blast loading, compared to the serviceability consideration that governs deflection limits in conventional design. All that is required from the design process is a demonstration that:

- the deformations do not cause collapse of any part of the structure
- no part of the structure impinges on critical equipment.

It should be noted that the deformation-limit approach permits the structure to form temporary mechanisms. These may occur when a series of plastic hinges forms during short duration, dynamic blast loads.

Deformation criteria for structural steelwork

The controlling criterion for the protection of buildings from blast is normally a limit on the deformation or deflection of the element. In this way, the degree of damage sustained by the element may be controlled. The damage level that may be tolerated in a particular situation will depend on what is to be protected.

Deformation criteria for beams can be defined in a variety of ways but the most common ways are by:

- ductility ratio
- support rotation.

Deformation in structural steel members is generally defined in terms of a ductility ratio, while support rotation limits are imposed for reinforced concrete elements (i.e. reinforced concrete floors).

For structural frames, the deformation criterion is based on defining limits to frame sidesway deflection and joint rotation.

Ductility ratio

The ductility ratio μ is defined by the relationship:

$$\mu = \frac{\text{total deflection}}{\text{deflection at the elastic limit}} = \frac{y_m}{y_{el}}$$

The ductility ratio is a measure of the ability of the structure to absorb energy. It is a property of the individual member or frame determined by topology and detailing.

TM 5-1300[12] recommends maximum values of ductility ratio to be used in structural steel design and these are presented in Table 10.1.

Table 10.1 *Allowable ductility ratios*

Component	Protection category	
	1	2
Steel beams	10	20
Reinforced concrete slabs	Not applicable	Not applicable

The protection categories are defined as:

Category 1   For the protection of personnel and equipment through the attenuation of blast pressures and to shield them from the effects of projectiles and falling portions of the structure.

Category 2   For the protection of the structural elements from collapse under the action of blast loading.

These ductility ratios are consistent with maintaining structural integrity into the plastic range. To justify the large hinge rotations assumed, it is essential that the yielded portions associated with the hinge are adequately restrained laterally to inhibit premature failure by instability.

Although these recommendations have been used extensively, some authors consider the ductility ratios to be high. Smith and Hetherington[5] also proposed ductility ratios based on a different acceptance criterion. In this case, acceptance criterion refers to reusable and non-reusable structures. These values are presented in Table 10.2.

Table 10.2 *Ductility ratios for design, proposed by Smith and Hetherington*

Ductility ratio μ	Design category
3	A reusable structure
6	A non-reusable structure

Support rotation

The deformation limits for reinforced concrete slabs are best expressed by maximum allowable support rotations. This method bears no direct relationship to the actual failure criteria of the structure, although there will be an approximate correlation for simple structures. The method may be useful in defining absolute limits for use in conjunction with other methods (see Figure 10.1).

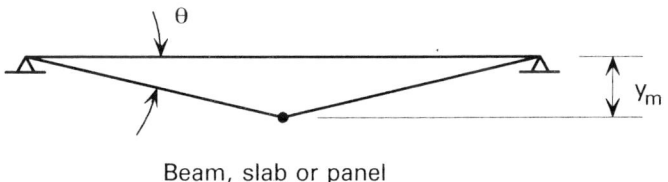

Beam, slab or panel

Figure 10.1 *Deformation limit based on member end rotation*

TM5-1300 recommends limits for support rotations in design based on the type of protection that is to be provided; these are presented in Table 10.3.

Table 10.3 *Allowable support rotations*

Component	Protection category	
	1	2
Reinforced concrete slabs/beams	2°	4°

The protection categories are as defined in the section on ductility ratios.

Deformation criteria for structural frames

For a frame structure, representation of the response by a single quantity has not generally been used. In lieu of a ductility ratio criterion, the amount of inelastic deformation is restricted by means of a limitation on the individual member rotation θ. The degree of sidesway that can be tolerated is a function of the potential P-Δ effects on the member and this ought to be assessed (see Figure 10.2).

For members that are not loaded between their ends (internal columns of buildings), θ is zero and only the sidesway criterion is to be considered. TM5-1300 limits the maximum member end rotation to 2° and maximum sidesway deflection to 1/25 of the storey height.

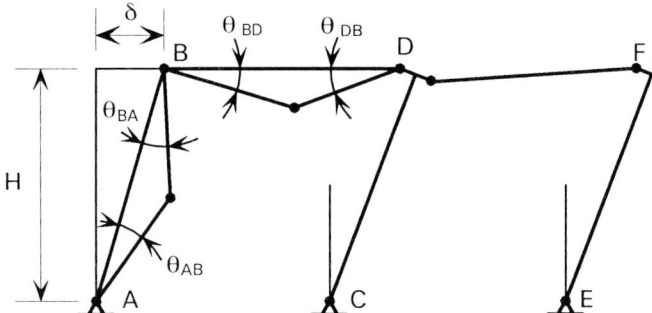

Figure 10.2 *Sidesway and member end rotations for frames*

10.3 Procedure to assess response and adequacy of structural components

A simple design procedure is presented showing how to determine the response and adequacy of an individual structural steel member subjected to blast loading. The response of the member is based on the single-degree-of-freedom analysis method.

1. Determine the blast load characteristics assuming that the blast load is triangular in profile and that the rise time is zero. This requires that the peak value of the load F and the duration of the load t_d is known (see Section 7).

2. Assume a loading/response regime, i.e. impulsive or dynamic/quasi-static. As a first guess, assume that dynamic/quasi-static conditions prevail (typical for steel members due to their low mass) (see Section 8.1).

3. Determine dynamic material properties based on an estimate of the *time to reach yield* criterion (see Section 9.2).

4. Assume an acceptable response criterion based on a maximum allowable ductility ratio μ (see Section 10.2.2).

5. Estimate the maximum member resistance value R_m required based on the preliminary design, assuming an equivalent static ultimate resistance based on the following table:

Structure type	R_m
Reusable	1.0F
Non-reusable	0.5-0.8F

   where F is the peak value of the load.

6. Select a steel section size that is not prone to lateral buckling. Refer to standard steel section tables and BS 5950-1.

7. Determine the required value of the plastic moment of resistance M_p. To determine M_p, a section modulus value is required; refer to BS 5950-1.

8. Calculate the natural period of the structural member using transformation factors for mass, stiffness, and load from the relationship

$$T = 2\pi\sqrt{\frac{K_{LM}M}{K}}$$

See Section 8.3 for further details.

9. Determine the available ultimate resistance of the structural member from Tables C.1 to C.3 in Appendix C.

10. Calculate the ratios R_m/F and t_d/T and determine y_m/y_{el} from Figure C.1. For the member section selected in step 6, the ratio y_m/y_{el} is the required ductility ratio for the member selected under the applied blast loading.

11. Check that the ductility ratio demand computed in step 10 is acceptable to the criterion assumed in step 4. If not, a different section must be selected and the process repeated.

12. Obtain dynamic reactions from Tables C.1 to C.3 and determine shear stress acting on the web. If the shear stress is less than the allowable shear stress of the section then shear plus bending combination is satisfactory.

13. Check that lateral torsional buckling does not occur (BS 5950-1). Note that the plastic hinge compressive zone can be quite long.

14. Check that the connections at the ends of the beam are adequate.

15. Check that the correct loading/response regime has been chosen. If the quasi-static case is not appropriate, the procedure has to be repeated using impulsive conditions (see Section 8.1).

If all the checks are satisfied, the structural member is adequate, however it may be possible to reduce the section size if the response is not near the maximum ductility ratio allowed.

11 DETAIL DESIGN AND STRUCTURAL CONNECTIONS

The components of a building that need to be considered when resisting blast loading include beams, columns, bracing, floors, roof, walls, and stairs. This Section considers the design aspects.

11.1 Beams

Members acting primarily in bending may also carry significant axial load caused by the blast loading. The latter can have a significant influence on both the strength and stiffness in bending, compression being detrimental and tension generally being beneficial. Even where the changes in stiffness are not important, account should be taken of the moment resulting from the combination of axial load and lateral deflection, for both elastic and plastic analysis.

In certain situations, beams may be subjected to load reversal and rebound effects; as a result, they can lose integrity. Steel beams may in some cases require lateral restraints in order to prevent buckling and to promote adequate rotational capacity. Steel sections should be able to resist full plastic moment capacity and should therefore be compact or plastic (BS 5950).

The effects of overestimating rotational fixity will be to calculate higher reaction forces, lower deflections, and lighter sections. Conversely, underestimating rotational fixity will result in higher deflections, lower reaction forces, and heavier sections. It is generally conservative (i.e. safe) to underestimate fixity unless the magnitude of forces transferred into the main structure is of particular concern.

Under large deformations, if there is sufficient axial fixity then membrane action will be developed and this will provide load-carrying capacity above that offered by the beam alone. Such additional capacity should not be relied on, however, unless the axial stiffness of the surrounding structure (including the beam connections) is sufficient. It is conservative to ignore the potential benefits of axial restraint for beam sizing.

11.2 Columns

Columns are predominantly loaded with axial forces under normal loading conditions, however under blast loading they may be subjected to bending and projectile loads. Such forces can lead to the loss of load-carrying capacity of a section. It is vital that in critical areas, full moment-resisting connections are made in order to ensure the load-carrying capacity of a column after an explosion.

Steel and reinforced concrete columns should be designed to resist such loading. Splices in vertical reinforcement should be designed to achieve full tension anchorage, and surrounding links should be anchored firmly. In critical areas, steel columns may benefit from localised concrete encasement, which increases the mass and strength, however good detailing and selection of an appropriate member size will be equally good.

The casing of ground floor perimeter steel beams and stanchions in reinforced concrete is beneficial in increasing their strength and stability. To some extent, such casing will compensate for any lack of connection continuity. When casing is added, it will permit any infill masonry to bed to the surrounding solidly. In turn, this will promote arching action within the masonry under the extreme condition of explosion damage, and it will promote the maintenance of fire resistance. A potential drawback is the risk of concrete spalling during the explosion, with the fragments acting as high-velocity projectiles.

11.3 Floors

Floors are also likely to be subjected to load reversal conditions. Floors not designed to resist upward forces may be overstressed or dislodged. It is therefore recommended that all floors in the immediate vicinity of an explosion (e.g. ground and first floors) should be fully restrained to the structure. Designers should therefore be careful that the flooring system does not attract unforeseen loads during an explosion.

Cast in-situ reinforced concrete floors are best suited to high overpressure loads as long as reinforcement is provided for uplift forces. To cater for this, reinforcement should be provided in both the *top* and *bottom* of the slab and, ideally, reinforcement should span in two directions.

Composite floors also require additional reinforcement in order to ensure that uplift forces do not cause concrete overstressing. Additional shear keys may be required in order to accommodate the increase in longitudinal shear forces. It is possible that steel sheet buckling may occur and reduce the ultimate load-carrying capacity of the floor.

Precast concrete floors are generally not suited to withstanding blast loads. They can be dislodged easily and cause further hazards by falling debris. Performance can be increased by tying them back to their supporting beams, using a dense screed with mesh reinforcement and using longitudinal ties.

11.4 Roofs

Roofs can be subjected to either severe compressive forces caused by external explosions, or may be subjected to uplift forces caused by internal explosions. Lightweight and glazed roofs should therefore be avoided where possible. Reinforced concrete or composite roofs can provide better blast response characteristics.

11.5 Walls

The load-carrying capacity of structural walls can be reduced as a consequence of the out-of-plane loading caused by blast loads. Walls should therefore be designed to behave in a ductile fashion by means of steel reinforcement. Spalling of concrete from the rear face of walls can be of concern as high velocity projectiles can be created in certain instances. In critical areas (such as temporary refuge areas), walls should be designed to resist such loads. The walls and their connections (i.e. ground and roof) should be designed for the increase in horizontal shear and bending from the applied blast pressures.

11.6 Cladding

Many proprietary forms of cladding are used to face buildings, the most common materials being stone, steel, or glazing. More often than not, the cladding units are prefabricated off-site and installed using fast-track steel fixing systems, which minimise the time spent on site. Due to prefabrication, and the requirement of fast-track installation, these fixing systems may not be easily accessible, and as a result can be difficult to inspect after an explosion.

Damage to the cladding fixings can occur in areas remote from the explosion source, despite there being no apparent damage in comparable areas subject to the full brunt of the explosion. Although the cladding panels may look undamaged (the material has the ability to absorb the relatively short-term loading and dissipate the applied energy by its own deflection), it is the integrity of the fixing system that may be critical.

Damage to fixing systems is most likely to be caused by sway or whiplash of the building frame. In these situations, the energy is transmitted directly into the fixings, and may cause fracture of the supporting elements or a loss in their anchorage capacity. Where a loss in anchorage capacity occurs, no indication of the loss of integrity may be visible. It is therefore imperative that demountability of panels, access, and inspection of fixings are considered in the design of the cladding. If not, a great deal of careful and highly-skilled examination will be required to ascertain the nature and extent of damage following an explosion.

11.7 Stairs

Stairs are often designed specifically for gravity loads, however blast loads can cause an uplift on stairs so they should be tied down to the structure.

11.8 Beam-column connections

Particular attention needs to be given to connection design for blast enhanced structures, for the following reasons:

- Connection forces, relative to the size of the member being connected, will be found to be very large.

- Where the member becomes plastic, the connection forces will be a function of the yield stress of the member (and this may be enhanced further by strain-rate effects).

- Force components not met in conventional design could be significant. For example, a beam that is subject to large deformations will resist the load partly by axial or membrane tension.

- There are high strain rates resulting in increased risk of brittle fracture. Good welding procedures are required if fractures are to be avoided during overload; see review of connection performance under overload[59].

11.8.1 Structural steel connections

Where possible, frame connections should provide full continuity, but all connections, whether fully rigid or nominally pinned, should exhibit ductility when overloaded. Great attention should be paid to their detailing, recognising their vital structural function. It must be recognised that the joints will be tested to their limit, without safety margin, under blast loading. This has to be considered within the knowledge that the connection form adopted contributes significantly to frame cost.

Where bolted connections are used and plastic hinges are expected to form, the capacity of the bolt arrangement should be greater than the plastic moment of the attached beam. The plastic moment of the beam should be an upper bound value taking account of yield variation and strain rates.

For fixed-ended connections, both maximum moment and maximum shear occur at the same point. It is necessary to demonstrate that the section can withstand the combined forces if necessary by plastic deformation, always ensuring that premature collapse by local or lateral buckling is prevented. Large rotations may be experienced at the same location and rotational capacity should be assured.

Particular care should be taken to ensure that steel connections can withstand both the dynamic loads and any load reversals imposed upon them. They should also be capable of accepting distortion without fracture. For simple shear connections that are nominally pinned, this implies detailing in such a manner that imposed distortion does not precipitate fracture. Instead, the various components are proportioned to carry reversal of loading. This may be important in extended end-plate design where usually the bolts are arranged efficiently only to carry gravity loading.

For connections that may be weak under imposed distortion, there are benefits in providing a second line of defence. It has been suggested in engineering journals that the use of a seating cleat in conjunction with bolted end plates is a preferred solution. Figure 11.1 shows an example of such a detail. The seating cleat is positioned 5-10 mm below the lower beam flange and serves to hold the beam in the event of any failure of the primary joint. It serves no function in normal operations. The problem with end plates of the type shown lies in the spacing of the bolts in relation to the web and the top flange. If this spacing is close, then under excessive imposed rotation there will not be enough end-plate material to bend and a fracture may be initiated along the web line.

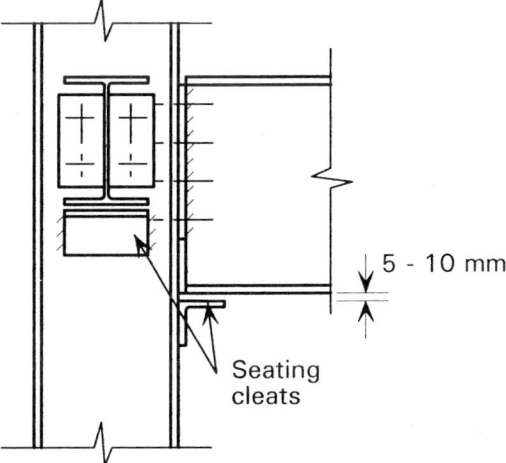

Figure 11.1 *Enhanced design of steel beam-column end plate connection*

The advantages of using such a detail are numerous:
- low fabrication costs (as can be shop welded)
- they provide a secondary means of stopping beam/floor collapse and thus preventing an escalation of events
- can aid in site erection.

The main drawbacks are:
- should not be used to carry loads during normal service life
- may interfere with architectural details
- provide only temporary support after an explosion.

12 FOUNDATIONS

For a structure to exhibit any measure of blast resistance, its frame and foundation must be capable of sustaining the large lateral loading. The design of foundations resisting blast loads introduces problems that are as yet unanswered in civil engineering. This is because there has been very little work done on dynamically-loaded footings and other foundations, and current practices are limited to design for static loading. Use of maximum blast overpressures to estimate static loads will probably result in grossly over-designed foundations. There is considerable evidence from explosion accident investigations that the foundations rarely fail, even for strong structures that have essentially been reduced to rubble by air blast. In many cases, because the dynamic effects from the blast loading are of such short duration, buildings do not slide laterally or rotate to any significant degree. Problems may be encountered, however, if loosely-compacted soils are present. Due to the dynamic effects, loose soils may be liquified or compacted, allowing structures to rotate or settle.

Only tentative guidelines for foundation design can be given because the design of a foundation is subject to uncertainties:

- the dynamic properties of soils (soils are strain dependent making the analysis nonlinear)
- the dynamic problem is sufficiently complex that extensive calculations are required, making it almost an impractical design procedure.

Owing to these uncertainties, only simplistic quasi-static analyses can be performed.

To design a foundation by a conservative quasi-static approach such that lateral motion will be small, it is recommended that the resistance available from the pressure at any time be made equal to or greater than twice the difference between the total lateral load on the footing and the available friction. In many foundation designs, it will be found uneconomical to prevent lateral motion of the structure by providing friction and passive pressures to resist the total lateral dynamic reactions on the footings.

13 NON-STRUCTURAL ENHANCEMENTS

Non-structural enhancements can provide significant protection to the building and its occupants. Three non-structural enhancements are considered: blast-enhanced glazing, facade detailing, and building internal layout.

13.1 Blast-enhanced glazing

Blast-enhanced glazing can be provided in buildings to reduce greatly the number of sharp-edged fragments that are created when ordinary annealed or toughened glass is subjected to blast. These shards, which travel at high speed, can cause severe injuries to personnel, damage delicate information technology equipment, and cause problems if they enter into air-conditioning systems. Glazing can be blast enhanced by:

- applying transparent polyester anti-shatter film (ASF) to the inner surface of the glazing
- applying transparent polyester anti-shatter film to the inner surface of the glazing with bomb blast net curtains (BBNC)
- glazing or re-glazing with high-strength glass that offers good blast resistance
- installing blast-enhanced secondary glazing inside the exterior glazing.

The modes of failure of the blast-enhanced glazing are such that the number of loose shards produced is decreased greatly (by over 90%) and the amount of glass falling from the building after the event is also reduced greatly. This enables access to the building to be significantly quicker and the clear-up is easier and quicker. Re-occupation can then be undertaken more promptly than in the case where glazing protection has not been installed (Figure 13.1).

Figure 13.1 *Blast-enhanced glazing providing protection [Courtesy of TPS Consult]*

Transparent anti-shatter film

The most easily applied anti-shatter film is a clear polyester film that is applied with a pressure-sensitive adhesive to the inside of glazed areas. The anti-blast protective qualities are provided partly by the strength, ductility, and thickness of the base film and significantly by the properties and thickness of the adhesive. Anti-shatter film can be factory applied before installation or afterwards as a retrofit measure.

In some circumstances, it is impractical to apply anti-shatter film, e.g. to glass with a textured surface, small intricately-shaped panes, overhead roof lights, and windows with internally-fixed security bars. In these and similar situations, a brush-applied liquid anti-shatter film can be used.

Bomb blast net curtains

An additional measure of protection against flying splinters from glass following an explosion can be provided by the installation of weighted net curtains in combination with anti-shatter film on the glass.

High-strength glass

High-strength glass consists of laminated annealed glass, laminated toughened glass, or polycarbonate glass. In double-glazed units, combinations of toughened and laminated glass can be used, the laminated glass generally being the inner pane, but preferably used for both panes.

Where high-strength glass is to be used, the frames and fixings must be strengthened to resist the additional load.

Secondary glazing

Blast-enhanced secondary glazing can be installed inside the exterior glazing. The secondary glazing prevents glass shards from entering the inside of the building.

Further information on detailed aspects of providing glazing protection for buildings is presented in Appendix D.

13.2 Facade detailing

There are several aspects of facade design that should be considered when attempting to minimise the vulnerability of people within the building and damage to the building itself, including:

- Minimising the amount of glazing in the facade. This limits the amount of internal damage from the glazing and the amount of blast that can enter and cause damage to the internal fixtures and fittings.

- Ensuring that the cladding is fixed securely to the structure with easily-accessible fixings. This will allow rapid inspection and, if necessary, replacement after an event.

- Ensuring that the cladding system allows for easy removal and installation of any one panel. This will prevent the necessity of removing all the panels if only one is damaged after the event.

- Avoiding the use of deep reveals and flat window sills that are accessible from ground level, as these provide ideal concealment places close to the building for the small, carried device.

- Minimising the use of deep surface modelling, as this can enhance the blast effects by reflection of the blast wave, and thus cause disproportionately more damage than would be the case with a plane facade.

- Unsecured objects attached to the structure should be tied to prevent them from falling and causing injury.

13.3 Internal layout of building

Internal layout planning should be undertaken with the aim of isolating *value* from the threat. The internal layout of the building should be arranged so that the highest exterior threat is separated by the greatest distance from the highest value asset. Other assets of lower value, in descending scale, are put in between to act as sacrificial layers of decreasing value, thereby achieving *quasi-stand-off*. It should be appreciated that stand-off can be achieved vertically as well as horizontally. An example of an internal layout arrangement emphasising this philosophy is shown in Figure 13.1.

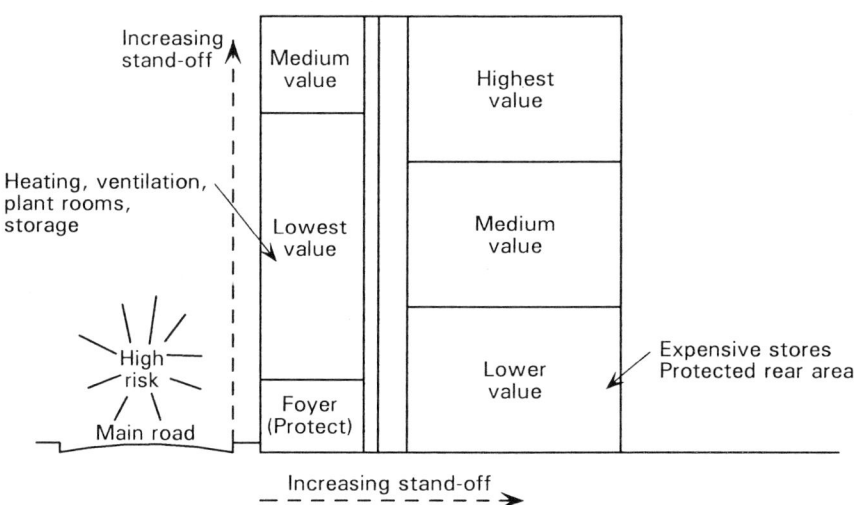

Figure 13.2 *Internal planning of a building*

14 INSPECTION OF DAMAGED BUILDINGS

Inspection of damaged buildings requires a coordinated effort of professionals (structural engineers, architects, building services, and quantity surveyors) and insurance bodies (loss adjusters).

An initial inspection is usually performed shortly after the explosion to identify the damage and propose an interim course of action. A second, more detailed investigation usually follows to ascertain damage more comprehensively and to make recommendations for the repair or rebuilding of the structure. These will cover both structural and non-structural (e.g. cladding) components of the building.

Particular attention must be paid to details that are hidden from view (e.g. cladding fixings) and to the definition of the area of concern. In certain cases, damage has been reported at a distance that was initially considered safe from damage.

The inspection of bomb-damaged buildings is a dangerous task. Full safety precautions should always be taken.

Once these investigations are completed, preliminary and full reports will be produced, presenting technical and cost information for the reinstatement of the building.

More detailed information on the inspection of damaged buildings can be obtained from the Institution of Structural Engineers publication *The structural engineer's response to explosion damage*[60].

REFERENCES

1. The Management of Health and Safety at Work Regulations
   Statutory Instrument SI 1992, No. 2051
   HMSO, London, 1992

2. BUILDING TECHNICAL FILE
   The Construction Products Directive of the European Communities, 89/106/EEC
   The Builder Group, London, 1989

3. The Construction (Design and Management) Regulations 1994 (Condam Regulations)
   Statutory Instrument SI 1994, No. 3140
   HMSO, London, 1995

4. Bombs - protecting people and property:
   A handbook for managers and security officers
   Home Office, March 1994

5. SMITH, P.D. and HETHERINGTON, J.G.
   Blast and ballistic loading of structures
   Butterworth Heinemann, 1994

6. THE BUILDING REGULATIONS
   Manual to the Building Regulations
   Department of the Environment and the Welsh Office, HMSO, 1991
   Approved Document A
   The Building Standards (Scotland) Regulations
   Technical Standards, Part D
   The Scottish Office, Edinburgh, HMSO, 1990
   The Building Regulations (Northern Ireland)
   Technical Booklet E
   The Department of Environment for Northern Ireland, HMSO, 1994

7. BRITISH STANDARDS INSTITUTION
   BS 5628: Code of practice for use of masonry
   BSI, 1985

8. BRITISH STANDARDS INSTITUTION
   BS 5950: Structural use of steelwork in buildings
   BSI, 1990

9. BRITISH STANDARDS INSTITUTION
   BS 8110: Structural use of concrete
   BSI, 1997

10. BRITISH STANDARDS INSTITUTION
    European Prestandard ENV 1991-2-7: 1998
    Eurocode 1: Basis of design and actions on structures,
    Part 2-7: Accidental actions due to impact and explosion
    BSI, 1998

11. AMERICAN SOCIETY OF CIVIL ENGINEERS
Design of structures to resist nuclear weapons effects, Manual 42
Washington, DC, 1985

12. US DEPARTMENTS OF THE ARMY, NAVY AND AIRFORCE
Technical Manual, Army TM5-1300, Navy NAVFAC P-397,
Air Force AFR 88-22
Structures to resist the effects of accidental explosions
US Department of Commerce, National Technical Information Service,
Washington, DC, 1990

13. US DEPARTMENT OF ENERGY
A manual for the prediction of blast and fragment loadings on structures:
DOE/TIC-11268
Washington, DC, 1992

14. DEFENSE NUCLEAR AGENCY
The design and analysis of hardened structures to conventional weapons effects
DAHS CWE, 1995

15. AMERICAN SOCIETY OF CIVIL ENGINEERS
Structural design for physical security - State of the practice report
Washington, DC, 1995

16. US DEPARTMENT OF THE ARMY
Technical Manual, TM5-855-1 Fundamentals of protective design for conventional weapons
US Department of Commerce, Washington, DC, 1986

17. DRAKE, J.L., TWISDALE, L.A, FRANK, R.A., DASS W.C., ROCHEFORT, M.A., BRITT, J.R., MURPHY, C.E., SLAWSON, T.R. and SUES, R.H.
ESL-TR-87-57 Protective construction design manual
Air Force Engineering and Services Centre, Tyndale Air Force Base, Florida, 1989

18. CHRISTOPHERSON, D.G.
Structural defence
Ministry of Home Security, London, RC450, 1946

19. RHODES, P.S.
The structural assessment of buildings subjected to bomb damage
The Structural Engineer, Vol. 52, No. 9, pp. 329-339, 1974

20. BAKER, W.E., COX, P.A., WESTINE, P.S., KULESZ, J.J. and STREHLOW, R.A.
Explosion hazards and evaluation
Elsevier, Amsterdam, 1983

21. BANGESH, M.Y.H.
Impact and explosion - Analysis and design
Blackwell Scientific Publications, Oxford, 1993

22. HOPKINSON, B.
    British Ordnance board minutes 13565, 1915

23. CRANZ, C.
    Lehrbuch der Ballistik
    Springer, Berlin, 1926

24. KINNEY, G.F. and GRAHAM, K.J.
    Explosive shocks in air, Second Edition
    Springer-Verlag Inc., New York, 1985

25. UHLENBECK, G.
    Diffraction of shock waves around various obstacles
    University of Michigan, Engineering Research Institute, Ann Arbor, Michigan, March 1950

26. BRITISH STANDARDS INSTITUTION
    BS 6399: Loadings for buildings
    Part 1: 1996: Code of practice for dead and imposed loads
    Part 2: 1997: Code of practice for wind loads
    BSI

27. NORRIS, C.H., HANSEN, R.J., HOLLEY, M.J., BIGGS, J.M., NAMYET, S. and MINAMI, J.K.
    Structural design for dynamic loads
    McGraw Hill, New York, 1959

28. BRITT, J.R., DRAKE, J.L., COBB M.B. and MOBLEY, J.P.
    BLASTIN user's manual ARA 5986-2 Contract DACA39-86-M-0213 for USAE Waterways Experiment Station, Vicksburg, Mississippi
    Applied Research Associates Inc., April 1986

29. BRITT, J.R., DRAKE, J.L., COBB, M.B. and MOBLEY, J.P.
    CHAMBER user's manual ARA 5986-1 Contract DACA39-86-M-0213 for USAE Waterways Experiment Station, Vicksburg, Mississippi
    Applied Research Associates Inc., April 1986

30. DYNA 3D User's Manual: Non-linear dynamic analysis of structures in three dimensions
    UICD-1952, Rev. 5
    Lawrence Livermore National Laboratory, Livermore, California

31. BRITT, J.R. and LUMSDEN, M.G.
    Internal blast and thermal environment from internal and external explosions: A user's guide for the BLASTX code, Version 3
    USACE Waterways Experiment Station, Vicksburg, Mississippi, 1994

32. McGLAUN, J.M., THOMPSON, S.L. and ELRICK, M.G.
    CTH. A three dimensional shock physics code
    International Journal of Impact Engineering, Vol. 10, pp. 351-360, 1990

33. BAUM, J.D., LUO, H. and LOHNER, R.
    Numerical simulation of blast in the World Trade Center
    Paper presented at the American Institute of Aeronautics and Astronautics 33rd Aerospace Sciences Meeting, Reno, Nevada, January 1994

34. WINDHAM, J.E., ZIMMERMAN, H.D. and WALKER, R.E.
    Improved shock predictions for fully buried conventional weapons
    Proceedings of Special Session of the Sixth International Symposium on the Interaction of Conventional Munitions with Protective Structures
    Panama City, Florida, 1993

35. GUNGER, M.
    Progress on tasks under the sympathetic detonation program
    WL/MN-TR-91-85
    Orlando Technology Inc., Shalimar, Florida, 1992

36. HIKIDA, S., BELL, R. and NEEDHAM, C.
    The SHARC Codes: Documentation and sample problems
    SSS-RS89-9878
    S-Cubed division of Maxwell Laboratories Distribution Ltd, Albuquerque, New Mexico, 1988

37. CONWEP
    Conventional weapons effects program
    Prepared by D.W. Hyde
    US Army Waterways Experiment Station, Vicksburg, Mississippi, 1991

38. KINERGY, C.N. and BULMASH, G.
    Airblast parameters from TNT spherical air burst and hemispherical air burst
    Technical Report ARBRL-TR-02555
    US Army Armament Research and Development Centre, Ballistics Research Laboratory, Aberdeen Proving Ground, Maryland, 1984

39. BIGGS, J.M.
    Introduction to structural dynamics
    McGraw-Hill, New York, 1964

40. BUDGE, K.G. and PERRY, J.S.
    AMMALE shock physics code written in C++
    International Journal of Impact Engineering, Vol. 14, pp. 107-120, 1993

41. BENSON and ASARO (eds)
    Computational methods for material modelling
    American Society of Mechanical Engineers Book No. H-00883, New York, 1993

42. LOHNER, R., YANG, C., CEBRAL, J., BAUM, J.D., LUO, H., PELESSONE, D. and CHARMAN, C.
    Fluid structure interaction using a loose coupling algorithm and adaptive unstructured grids
    Paper presented at the 26th American Institute of Aeronautics and Astronautics Fluid Dynamics Conference, San Diego, California, June 1995

43. SCHLAMP, R.J., HASSIG, P.J., NGUYEN, C.T., HATFIELD, D.W., HOOKHAM, P.A. and ROSENBLATT, M.
MAZe User's Manual
TRT Corporation, Los Angeles, California, 1995

44. THOMPSON, W.T.
Theory of vibrations with applications, Third Edition
Unwin Hyman, London, 1988

45. JARRETT, D.D.
Derivation of British explosives safety distances
Annals of the New York Academy of Sciences, Vol. 152, Art. 1, pp. 18-35, 1968

46. BOWEN, I.G., FLETCHER, E.R. and RICHMOND, D.R.
Estimate of man's tolerance to the direct effects of airblast
Technical Progress Report DASA-2113, Defense Atomic Support Agency, US Department of Defense, Washington, DC, 1968

47. AHLERS, E.B.
Fragment hazard study
Minutes of the 11th Explosives Seminar, Vol. 1, Armed Services Explosives Safety Board, Washington, DC, 1969

48. BAKER, W.E., WESTINE, P.S., KULESZ, J.J., WILBECK, J.S. and COX, P.A.
A manual for the prediction of blast and fragment loading on structures: DOE/TIC-11268
US Department of Energy, Amarillo, Texas, 1980

49. ELLIOT, C.L. and MAYS, G.C.
The protection of buildings against terrorism and disorder
Discussion, ICE Proceedings, Structures and Buildings, Vol. 104, Issue 3, August 1994 (26784)

50. NEWMARK, N.M. and ROSENBLUETH, E.
Fundamentals of earthquake engineering
Prentice-Hall Inc., Englewood Cliffs, New Jersey, 1971

51. BRITISH STANDARDS INSTITUTION
BS EN 10025: Hot rolled products of non-alloy structural steels - Technical delivery conditions
BSI, 1993

52. BRITISH STANDARDS INSTITUTION
BS EN 10210: Hot finished structural hollow sections of non-alloy and fine grained structural steels
BSI, 1994

53. BRITISH STANDARDS INSTITUTION
BS 4449: Specification for carbon steel bars for the reinforcement of concrete
BSI, 1997

54. BRITISH STANDARDS INSTITUTION
    BS EN 10088: Stainless steels
    BSI, 1995

55. THE STEEL CONSTRUCTION INSTITUTE
    Fire and Blast (FABIG) Technical Note (to be published in 1999)
    Design of stainless steel blast walls
    Confidential to FABIG members

56. BRITISH STANDARDS INSTITUTION
    BS 5950: Structural use of steelwork in buildings
    Part 1: Code of practice for design in simple and continuous construction: Hot rolled sections
    BSI, 1990

57. HORDYK, M. and MALIK, A.S. (eds)
    Joints in simple construction - Vol. 1: Design methods, Second Edition
    The Steel Construction Institute and The British Constructional Steelwork Association, 1993

58. BRITISH STANDARDS INSTITUTION
    Draft document DC 98/102164
    Proposed revision of BS 5950: Structural use of steelwork in buildings
    Part 1
    BSI, 1998

59. Vulnerability of existing steel framed buildings following the 1994 Northridge (California, USA) earthquake: Considerations for their repair and strengthening
    P. Maranian Brandow & Johnston Associates, Los Angeles, California
    The Structural Engineer, Vol. 75, No. 10, 20 May 1997

60. THE INSTITUTION OF STRUCTURAL ENGINEERS
    The structural engineer's response to explosion damage
    ISE, 1995

61. Guidance on the use of anti shatter film and bomb blast net curtains
    The Security Facilities Executive (SAFE)
    The Special Services Group (SSG), London, 1993

62. Glazing enhancement to reduce hazard
    The Security Facilities Executive (SAFE)
    The Special Services Group (SSG), London, 1993

63. General specification for security grade anti shatter film
    The Security Facilities Executive (SAFE)
    SAFE document SSG/EP/23/95
    The Special Services Group (SSG), London, 1995

64. General specification for bomb blast net curtains
    The Security Facilities Executive (SAFE)
    SAFE Document SSG/EP/7/96
    The Special Services Group (SSG), London, 1996

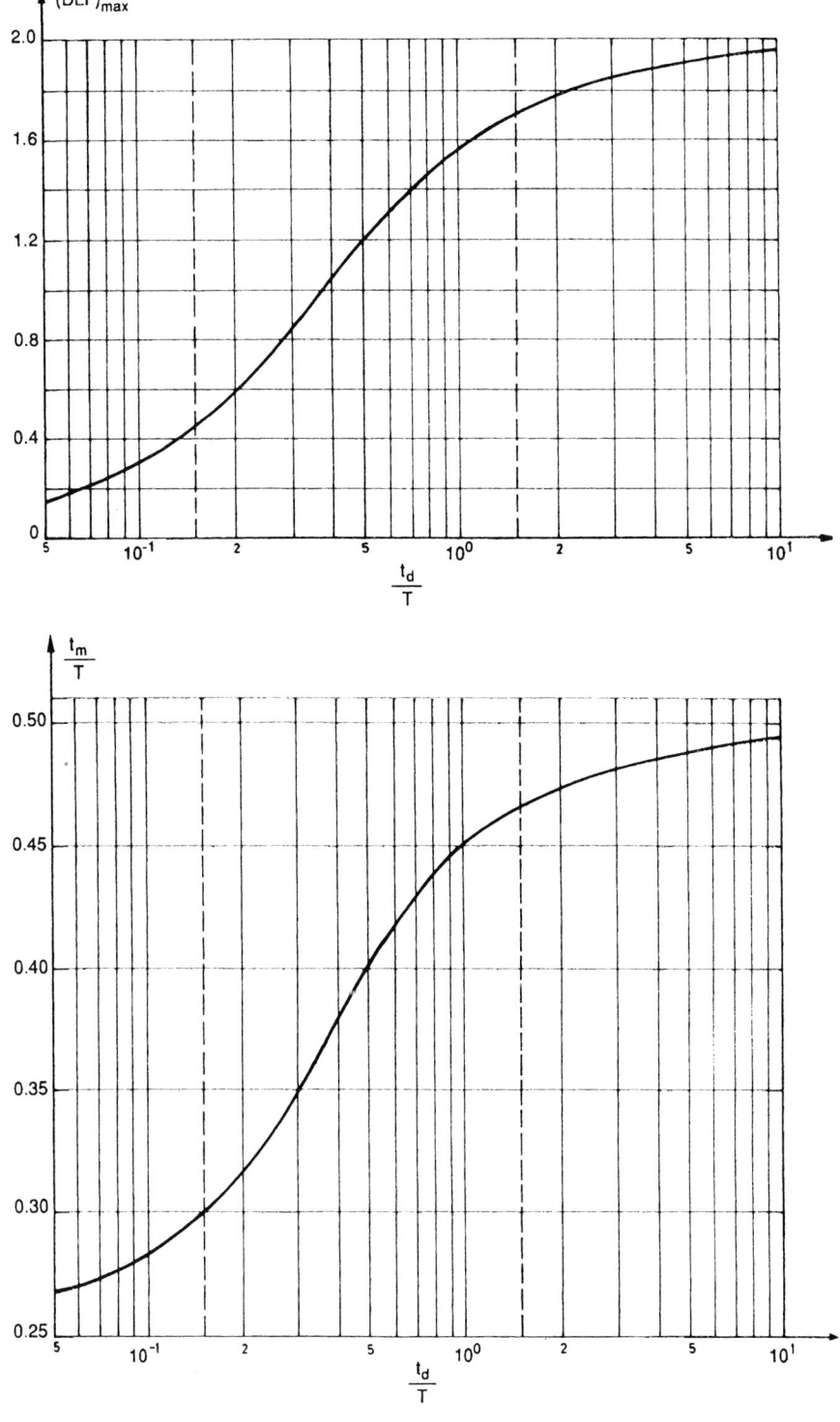

Figure C.3 *Maximum response of a single-degree-of-freedom elastic system due to a triangular pulse with zero rise time*

APPENDIX D DESIGN FOR GLAZING PROTECTION

Current policy requires that all government buildings should incorporate measures to mitigate explosion hazards, so documentation is available that assists designers to specify appropriate blast-resistant glazing systems for buildings. These documents are available from The Security Facilities Executive (SAFE), in particular the Special Services Group (SSG). The Security Facilities Executive is a UK Government Agency of the Cabinet Office (Office of Public Service) that provides security services to public clients. Within The Security Facilities Executive, the Special Services Group provides security-related technical research, development, advisory, and project operation services, within which the Explosives Protection team focuses on protective security measures against blast and weapons effects.

In this Appendix, further information pertaining to glazing protection is presented, based on the following documents from the Special Services Group:

- *Guidance on the use of anti shatter film and bomb blast net curtains*[61]
- *Glazing enhancement to reduce hazard*[62]
- *General specification for security grade anti shatter film*[63]
- *General specification for bomb blast net curtains*[64].

There are three main ways of providing glazing protection. They are:

- applying transparent polyester anti-shatter film (ASF) to the inner surface of the glazing, with or without the provision of bomb blast net curtains (BBNC)
- glazing or re-glazing with high-strength glass that offers good blast resistance
- installing blast-enhanced secondary glazing inside the exterior glazing.

Blast-enhanced glazing generally consists of laminated annealed glass or laminated toughened glass or, in double-glazed units, combinations of toughened and laminated glass, the laminated generally being the inner pane, but preferably both panes. Blast-resistant glazing will require suitably designed robust frames and fixings. Frames and fixings should be designed to fail before the cladding.

The primary purpose of glazing protection is to reduce the number of sharp-edged fragments that are created when ordinary annealed or toughened glass is subjected to blast. These shards, which travel at high speed, can cause severe injuries to personnel and can damage delicate information technology equipment and cause problems if they enter into air-conditioning systems.

Where glazing protection is provided, access to the building is made available sooner, allowing re-occupation of the building to be undertaken promptly.

D.1 Anti-shatter film (ASF)

Anti-shatter film can be applied to the inner surface of the glazing using an adhesive or liquid film, with or without the provision of bomb blast net curtains (BBNC).

D.1.1 Adhesive film

Adhesive anti-shatter film is a clear polyester film of a suitable specification at least 100 micron (0.1 mm) thick and certified at least Class B to BS 6206[65], applied with a pressure-sensitive adhesive to the inside of glazed areas. The anti-blast protective qualities are provided partly by the strength, ductility, and thickness of the base film and partly by the properties and thickness of the adhesive. In the event of an explosion, the shattered glass is held together by the film, so that dangerous shards do not fly off.

For windows over 3 m^2 in area or where the glass is over 6 mm thick, film having a BS 6206 Class A certificate should be used. This quality film should also be used if bomb blast net curtains are not installed. The use of 300 micron film should be considered for panes over 10 m^2 and for ground floor windows over 3 m^2 where bomb blast net curtains are not installed.

On new windows, or areas being re-glazed, the new glass should preferably be treated before fixing into the frames, with the film completely covering the glass to its extreme edges, while on existing windows the film must be applied as close as possible to the putty. The film cannot be applied to the patterned side of frosted, figured glass, although consideration may be given to reversing the glass so that the plain face is on the inside.

The proper application of anti-shatter film to double-glazed windows depends on the type of window concerned. If the window consists of two separate frames and, in normal use, the inner frame can be opened independently of the outer frame, both panes should be treated with film on the room side of the glass. If the inner pane cannot be opened independently, or *sealed-unit* double glazing is fitted, the inner pane must be treated. It should be noted that the inner pane of some double-glazing systems is only lightly fixed; if this is the case, it should be refitted more securely.

A peel-adhesion strength test has been developed for on-site quality control of the installation of film and for checking whether degradation of performance has taken place during its service life, which at present is between 5 and 10 years. The film is easily scratched or marked, so special care is required when cleaning. Standard anti-shatter film is designed for internal use only; if fixed externally, the film and adhesive will weather and degrade more quickly, more than halving its effective life.

D.1.2 Liquid film

In some circumstances, it is impractical to apply 100 micron anti-shatter film, i.e. to glass with a textured surface, small intricately-shaped panes, overhead roof lights, and windows with internally-fixed security bars. In these and similar situations, it is recommended that a brush-applied liquid anti-shatter film is used. Typically, a 350 micron coating thickness is applied to a clean and de-greased inside face of the glass.

D.1.3 Bomb blast net curtains (BBNC)

An additional measure of protection against flying splinters from glass following an explosion can be provided by the installation of weighted net curtains in combination with anti-shatter film on the glass. Following an external explosion that shatters the window, the curtain material *spinnakers* out and arrests the flying glass, which, because of the anti-shatter film, will be composed of larger pieces travelling at a slower velocity than would be the case if the glass were untreated. Under slightly more extreme conditions, the curtains will break away from the top fixings and at the same time billow outwards, wrapping up the glass and depositing it on the floor near the window.

Bomb blast net curtains are typically 90 or 100 denier warp knitted polyester terylene curtain material with a bursting strength of 500 kN/m^2. Curtains should cover the width of the window without a break, but if a break is necessary on wide windows, it should be located on the mullions. For the curtains to be most effective, they should have a weight box at the bottom that contains the excess weighted net curtain, i.e. the length of the curtain material is longer than the window.

Curtains should normally be installed close to the glass, typically 50 to 100 mm from the glass. This closeness is preferred because if the curtains are further away, detached shards may have the opportunity to twist and cut through the net, and blast waves passing around the glass may cause the curtains to billow away from the glass so that the shards are not effectively caught. Because the filmed glass loses velocity quite rapidly after the first few hundred millimetres, it is acceptable for the curtains to be fixed further back if other overriding conditions dictate.

Bomb blast net curtains should not be used where ordinary unprotected glass is present. If no anti-shatter film is present, the glass shards will not be contained by the curtain material.

D.1.4 Performance of anti-shatter film and bomb blast net curtains

An acceptable level of hazard mitigation for anti-shatter film and bomb blast net curtains has been confirmed by numerous blast tests covering various thicknesses of film and sizes of window. These have demonstrated a gain in protection as film adhesive and application control have improved since the 1970s, and also with the introduction of 100 micron film (as policy since 1985) instead of the former 50 micron film. With current on-site quality control assurance through peel adhesion testing, it can be reckoned that for normal modern windows at least 80% of the protection benefit is contributed by the film, with the curtains providing the remaining top-up protection. These proportions are approximate and depend on numerous factors including bomb size and stand-off, rebate and beading strength, pane size, and, perhaps most significantly, window design. For example, Georgian-style old windows with small panes and slim wood bars break up into many hazardous pieces that bomb blast net curtains can effectively contain.

For normal office-type single-glazed windows, the acceptable standard of hazard mitigation protection is deemed to be provided by anti-shatter film in conjunction with bomb blast net curtains, and for window glass up to 6 mm thick the anti-shatter film thickness should be 100 micron.

As a general rule, the specifications set out in Tables D.1 and D.2 can be used to provide sensible parity of protection against car bomb and van bomb threats to the occupants of multi-storey office blocks.

Table D.1 *Anti-shatter film and bomb blast net curtain recommendations for panes up to 3 m^2 and 6 mm thick*

Threat	Car bomb	Van bomb
Ground to 5th floors	100 ASF + BBNC	100 ASF + BBNC
6th to 11th floors	100 ASF only	100 ASF + BBNC
Above 11th floor	No protection	100 ASF only

Film identified as 100 ASF shall be at least 100 micron thick and should be certified to pass BS 6206 Class B.
Where bomb blast net curtains are omitted, Class A anti-shatter film should be substituted for 100 micron anti-shatter film except for very small and Georgian-style panes.

Table D.2 *Anti-shatter film and bomb blast net curtain recommendations for panes over 3 m^2 or over 6 mm thick*

Threat	Car bomb	Van bomb
Ground to 1st floors	Class A ASF + BBNC	Class A ASF + BBNC
2nd to 5th floors	Class A ASF only	Class A ASF only
6th to 11th floors	100 ASF only	Class A ASF only
Above 11th floor	No protection	100 ASF only

300 micron anti-shatter film should be considered at ground floor if bomb blast net curtains are omitted.

D.2 High-strength glazing

High-strength blast resistant glazing is available in numerous forms. The most common forms are toughened glass, polycarbonate, and laminated glass.

D.2.1 Toughened glass

Toughened glass can resist high blast pressures without damage, provided that it is well supported in a strong and rigid frame (though not requiring deep rebates as are necessary for laminated glass). Toughened glass is typically 3 to 5 times as strong as ordinary, plain, annealed glass, however it can be shattered unpredictably by small bomb fragments, and, thickness for thickness, is not as protective as well-held laminated glass. It has uses in certain situations and can survive undamaged at stand-off ranges that would cause plain glass to crack. Its use can therefore save large areas of glazing replacement after an incident. It is often recommended that anti-shatter glass be applied to the inner face of toughened glass to minimise the damage caused by flying fragments.

D.2.2 Polycarbonate

The primary use of polycarbonate as a blast-resistant material is normally as a secondary glazing barrier behind existing external glazing. Owing to the relatively high thermal coefficient of expansion, when polycarbonate is used in sheets exceeding 1 m by 1 m, special provision must be made for thermal movement in its frame.

D.2.3 Laminated glass

Laminated glass has a long life and can provide excellent resistance to blast provided that it is properly supported in robust frames with deep rebates. It is not normally necessary to apply anti-shatter film to laminated glass. Under the effect of remote blast, the shattered glass adheres well to the polyvinyl butyral interlayer. The addition of anti-shatter film can, however, be justified where close-in detonations or fragments (grenades, mortars, or nail bombs) are envisaged. Proprietary anti-spall systems exist that enhance the penetration resistance, the overall strength, and blast-resistant capacity of the laminated glass, and have protective value in appropriate situations. For new buildings and replacement windows, laminated glass should be considered in preference to anti-shatter film, for windows judged to be at risk.

The minimum overall thickness of laminated glass that should normally be used is 7.5 mm, including a minimum polyvinyl butyral layer thickness of 1.5 mm. Where 7.5 mm laminated glass has been specified to resist blast, the glass should be fixed in a robust, secure frame, which, together with its fixing, should be designed to withstand a static force of 7 kN/m^2 over the complete area of glazing and frame. For panes with a dimension of 1 m or more, a frame rebate of at least 35 mm should be provided to give 30 mm bearing. The pane should be bedded on both sides in a gasket or glazing material recommended by the manufacturer and secured internally by screwed beads. For smaller panes or purpose-designed blast-resistant frames with glazing edges bonded in adhesive sealant, the frame rebate for 7.5 mm glass may be reduced to 30 mm to give 25 mm bearing. Table D.3 shows the minimum glazing rebate for laminated glass. In all cases, care should be taken to eliminate unavoidable weaknesses; for example, weak internal snap-on beading should not be used, and additional screws and fixings should be provided where necessary.

Table D.3 *Frames and fixing design parameters for laminated glass*

Laminated glass thickness (mm)	Approximate glass pane size (m^2)	Equivalent ultimate static load (kN/m^2)	Minimum glazing rebate (mm)
6.4	0.6	6	25
	1.8	3	25
	3.0	3	30
6.8	0.6	8	25
	1.8	4	25
	3.0	4	30
7.5	0.6	12	25
	1.8	7	30
	3.0	6	30
11.5	0.6	18	25
	1.8	11	30
	3.0	9	30

D.2.4 Double-glazed units

Double-glazed panes can generally be considered to be less hazardous than single panes under blast. It may be justifiable to laminate the inner pane only but it is preferable to laminate both panes. The panes can be relatively thin (e.g. 6.4 mm + 6.4 mm) for hazard mitigation purposes. It should be noted that the inner pane of some double-glazing systems is only lightly fixed and if this is the case it should be secured.

An effective form of protection can be provided by a double-glazed unit where a toughened outer glass pane is combined with a laminated inner pane. A typical size is 6 mm toughened outer pane and 6.4 mm laminated inner pane, fixed securely in a standard frame but with enhanced frame fixings. This form of double glazing will offer a similar hazard reduction to anti-shatter film and bomb blast net curtains and is likely to be more cost-effective in the long term. The outer toughened pane will resist low blast loads without breaking, so there will be no damage to the glass and the inside of the building. Under large blast loads, the energy is absorbed in the outer pane breaking and, providing the inner pane is secured correctly, it can crack and absorb energy by yielding of the polyvinyl butyral inter-layer.

Specialist advice should always be sought where:

- glass thicker than 7.5 mm is to be used
- blast resistance in excess of 7 kN/m^2 is to be used
- recommended installation standards cannot be met
- specific threats are to be resisted.

Where only nominal hazard reduction is required (e.g. equivalent to that provided by anti-shatter film), thinner laminated glass and/or standard rebates may be acceptable. If the threat risk is considered low enough not to justify strong frames and deep rebates, it must be accepted that the laminated glass could blow into the room in much the same manner as glass protected with anti-shatter film. In such circumstances, it may be accepted that the potential blast resistance of 7.5 mm laminated glass will not be achieved, and the cost savings of thinner glass (say 6.8 mm) can be justified. Laminated glass in weak, shallow rebates cannot be considered significantly safer than anti-shatter film, but it is likely to hold the shattered glass together well, and bomb blast net curtains are not normally required.

Laminated glass in normal window frames is safer than anti-shatter film and bomb blast net curtains, subject to frame edge retention. It is also recommended in lieu of anti-shatter film and bomb blast net curtains for new build or when replacing windows.

D.2.5 Other materials

Toughened laminated glass and laminated glass/polycarbonate composites have excellent protective qualities against blast, however their use to date has not been extensive and specialist advice should be obtained regarding the use and framing designs of these materials. Reference may be made to BS 6206, BS 952[66], and BS 6262[67] for more information on any of the glazing mentioned above.

D.2.6 Wired glass

Georgian wired glass and other similar products are not blast proof and do not possess significant resistance to fragmentation. When subjected to a blast, glass fragments are produced in addition to the metal fragments, both of which may cause serious injury to personnel and damage internal office equipment.

D.3 Protection versus cost of glazing units

The type of glazing that can provide adequate protection for a particular threat (based on the assumed size of the device and the stand-off distance) can be obtained using the information contained in Tables D.4 and D.5.

Table D.4 *Approximate stand-off distances to produce internal flying glass*[68]

Device	Stand-off for 4 mm annealed glass (m)*
Small package	10
Small briefcase	14
Large briefcase	20
Suitcase	26
Car	60
Small van	120
Large van	140
Small lorry	160
Large lorry	200

* These distances can be changed significantly if local concentration of blast effects occurs.

Knowing the stand-off distance determined in the threat assessment and the stand-off distance where internal flying glass is produced for 4 mm annealed glass, the ratio

$$\frac{\text{stand-off distance for 4 mm annealed glass}}{\text{stand-off distance assumed in threat assessment}}$$

can be determined for the device identified in the threat assessment from Table D.4. This ratio is termed the *Comparative value*.

Knowing the Comparative value, the type of glazing that is adequate for the threat considered can be obtained directly from Table D.5.

Table D.5 *Comparison of glazing protection and relative cost*

Glazing type	Comparative value	Cost
Annealed glass		
4 mm	1.0	1.0
4 mm plus ASF	1.7	1.5
4 mm plus ASF and BBNC	2.0	1.8
Toughened glass		
6 mm	2.0	1.6
6 mm plus ASF	2.5	2.1
8 mm	2.5	2.4
8 mm plus ASF	2.9	2.9
10 mm	2.9	2.9
10 mm plus ASF	3.3	3.4
Laminated glass		
6.4 or 6.8 mm	2.5	2.1
7.5 mm	2.9	3.2
11.5 mm	3.3	3.9
Double-glazed units		
6 mm T plus 6 mm T	2.5	3.2
6 mm A plus 7.5 mm L	3.3	4.6
6 mm T plus 7.5 mm L	4.0	4.8

Having chosen a particular glazing type, an indication of the cost compared to 4 mm annealed glass can also be obtained. A number greater than 1 indicates a greater cost. Tables D.4 and D.5 should, however, be used with caution and only as an indicator.